Studies in the Psychosocial

Series Editors

Stephen Frosh
Department of Psychosocial Studies
Birkbeck, University of London
London, United Kingdom

Peter Redman
Department of Social Sciences
The Open University
Milton Keynes, United Kingdom

Wendy Hollway
The Open University
Hebden Bridge, West Yorkshire, United Kingdom

Psychosocial Studies seeks to investigate the ways in which psychic and social processes demand to be understood as always implicated in each other, as mutually constitutive, co-produced, or abstracted levels of a single dialectical process. As such it can be understood as an interdisciplinary field in search of transdisciplinary objects of knowledge. Psychosocial Studies is also distinguished by its emphasis on affect, the irrational and unconscious processes, often, but not necessarily, understood psychoanalytically. Studies in the Psychosocial aims to foster the development of this field by publishing high quality and innovative monographs and edited collections. The series welcomes submissions from a range of theoretical perspectives and disciplinary orientations, including sociology, social and critical psychology, political science, postcolonial studies, feminist studies, queer studies, management and organization studies, cultural and media studies and psychoanalysis. However, in keeping with the inter- or transdisciplinary character of psychosocial analysis, books in the series will generally pass beyond their points of origin to generate concepts, understandings and forms of investigation that are distinctively psychosocial in character.

More information about this series at
http://www.springer.com/series/14464

Matthew Adams

Ecological Crisis, Sustainability and the Psychosocial Subject

Beyond Behaviour Change

palgrave
macmillan

Matthew Adams
School of Applied Social Science
University of Brighton
Brighton, United Kingdom

Studies in the Psychosocial
ISBN 978-1-137-35159-3   ISBN 978-1-137-35160-9   (eBook)
DOI 10.1057/978-1-137-35160-9

Library of Congress Control Number: 2016956236

Cover illustration: © Rebecca van Ommen / Getty Images

Printed on acid-free paper

This Palgrave Macmillan imprint is published by Springer Nature
The registered company is Macmillan Publishers Ltd.
The registered company address is: The Campus, 4 Crinan Street, London, N1 9XW, United Kingdom

For Clare. Now will you make me a cup of tea? For Dylan, Amelie and Leila, thankfully and reliably indifferent to me writing a book ('is it a story?').

Our own chatter erupts in response to the abundant articulations of the world: human speech is simply our part of a much broader conversation.

David Abram

Acknowledgements

Thanks to Wendy Hollway for her supportive and constructive feedback on earlier drafts of this book. Thanks also to my colleagues in the School of Applied Social Sciences at the University of Brighton. I am grateful to be working in such a convivial place, where critical and interdisciplinary research is encouraged and nurtured. Thanks to Viv, Steve, Sam and Eve, always. Thanks to David Kidner, for inspiring me to write this book in the first place; and to Nigel Edley, Martin Jordan and Paul Stenner for offering ideas and encouragement along the way.

Contents

1

Introduction: The Walls Are Closing In

Introduction

The image that keeps coming to my mind is a nightmarish one inspired by Edgar Allen Poe. We are all in a room with four walls, a floor, a ceiling and no windows or door. The room is furnished and some of us are sitting comfortably, others most definitely are not. The walls are advancing inwards gradually… making us all more uncomfortable, advancing all the time, threatening to crush us to death. There are discussions within the room, but they are mostly about how to arrange the furniture. People do not seem to see the walls advancing. (Holloway 2010, p. 8)

'The walls closing in' is in some sense an apt analogy for the contemporary convergence of ecological crises, and our seeming inability to avert our predicament. John Holloway uses the analogy, in the passage cited here, to convey the apparently 'unstoppable advance of capital' as it pushes us ever closer to ecological destruction (Holloway 2010, pp. 8–9). It also captures the conjunction of a broader set of drivers of unfolding environmental crises, in evocative temporal and spatial dimensions. Global temperatures, peak oil predictions, species extinctions, deforesta-

© The Author(s) 2016
M. Adams, *Ecological Crisis, Sustainability and the Psychosocial Subject*,
DOI 10.1057/978-1-137-35160-9_1

..., glacier melting, desertification and the increased concentration of carbon dioxide in the atmosphere, for example, are often framed and animated in terms of what is already experienced as loss for some (Gupta 2011), and what further losses are to come, often with precise timescales, associated calculations of harms, multiple projected scenarios of scarcity and related problems for humanity.[1] For those lucky enough to be not, as yet, directly and detrimentally experiencing the effects of environmental crises, their reality, in other places or near futures, is communicated with mounting sophistication and urgency—the walls are closing in.

Consequently, there is a need for change and therefore action, commonly formulated as a need for 'sustainable development'. Despite the extensive debate over the specific nature of unsustainable human behaviour, there is little doubt that consumption practices are central. As such, consumption is seen as a primary site of responsibilization and potential transformation. In line with this logic, 'change-agents' combine information with attempts to persuade a significant proportion of the populations of wealthy nations to modify their consumption habits (Mazur 2011; Monroe 2003; Moser and Dilling 2006).

The first step in this process might reasonably be considered to be raising awareness and sharing concern—to get us to notice that the walls *are* closing in. This is the motivation behind information campaigns, the regular IPCC reports, increasingly sophisticated forms of communication, and to date, most policy approaches devoted to encouraging sustainable behaviour. But as the last two sentences of Holloway's quote indicate, there is a common perception that our response to date has been wholly insufficient. Inaction tends to be formulated as resistance to significantly shift consumption habits, but we might add, following Norgaard, the lack of a broader and sustained 'public response', that includes 'social movement activity, behavioural changes, or public pressure on governments' (Norgaard 2006, p. 373).

The lack of change is articulated as unsurprising by an emerging 'critical consensus' in the social sciences (Adams 2012). Theory and research on sustainable behaviour in the social sciences increasingly questions individualistic models of behaviour change, emphasizing, for example, the causal role attitudes, motivations or norms play in behaviour (e.g. Webb 2012). The infamous 'gap' between values and actions, or attitudes and behaviour, is the widely shared basis for this critique, here identified by Warde as the 'puzzle... that, if asked, almost everyone is in favour of protecting and preserving the environment and slowing climate change... However, people do not act consistently on the basis of their declared values and best intentions' (Warde 2013, p. 2) Despite the meaningful persistence of scepticism in some quarters, environmental destruction is increasingly accepted as a reality, as a consequence of human behaviour, and as a phenomena which is significantly impacting, and will have further impact on human and nonhuman life.[2] The salient feature of a range of critiques here is the claim 'that social context has been undertheorized and researched in psychological work on environmental problems and solutions, resulting in individualistic understanding of human practices and misguided policy interventions assuming people behave as rational agents' (Adams 2012, p. 217; Szerszynski and Urry 2010; Uzzell and Räthzel 2009; Webb 2010).

How we might productively cultivate this insight is more contested. Subsequent developments take off in various disciplinary directions, taking in psychology, social sciences and the humanities, anthropology and numerous interdisciplinary endeavours with arts and natural sciences. The focus of this book is a number of these strands, with a specific emphasis on the psychosocial. What does it mean to approach the issue of ecological crisis and sustainability psychosocially? The modest hope of this book is to bear witness to the ways in which human beings embody, and are embedded in, the social, material and cultural infrastructure at the heart of anthropogenic ecocide. To do so requires an interrogation of societal structures *and* their subjective dimensions, often lurking in the background as given in 'weak' approaches sustainability (Räthzel and Uzzell 2009).

In the chapters that follow, these dimensions intersect to the extent that they are often purposefully dissolved—into units of analysis that are conceptualized as hybrids of both, articulating the *in between*. This is the basis for what I consider to be a psychosocial approach, my understanding of which is indebted to Paul Stenner's work. He defines the psychosocial as 'an approach which attends to experience as it unfolds in and informs those networks and regimes of social interactivity (practice and communication) that constitute concrete historical and cultural settings' (2014, p. 205). Like Stenner, I adhere to an open, non-foundational definition of the psychosocial, and avoid collapsing it into a version of (social) psychoanalysis, which has sometimes been the case in the history of the term's use. I am interested in developing an account which emphasizes the 'often unrecognized, vague and fuzzy spaces inbetween forms of reality, knowledge and practice' (Brown and Stenner 2009, p. 49). I am unapologetically drawn to those perspectives that recognize this vital dimension and avoid the more obvious traps of psychological reductionism and social determinism in doing so. Echoing Stenner again,

> a psychosocial approach is valuable because questions of psychology can be very poorly posed when abstracted from their cultural, societal and historical settings, and likewise, because these settings are poorly understood in abstraction from the living, experiencing human beings whose actions make their reproduction and transformation possible. (Stenner 2014, p. 206)

Understanding human behaviour in an unprecedented era of anthropogenic ecological crisis demands this level of explanation. It is the only hope we have for making sense of how humans comprehend and respond to an ecological crisis that reflexively embeds them in it.

Chapter Guide

Chapter 2 first provides a brief overview of the ecological crisis as it is commonly understood by an international scientific consensus. It adopts a critical realist standpoint in relation to anthropogenic ecological degradation, and this position is explained, with reference to the geological concept of the Anthropocene. The second focus of this chapter is to reflect on how ecological degradation is communicated. A key element of this communication is of course information about the role of human behaviour in creating and maintaining ecological degradation. Science communicators have often worked on the basis of an information-deficit model, with the implication that more and better information is needed to provoke more sustainable behaviour. To date, however, the impact of information and its communication, whether measured in terms of changes in individual consumption patterns or a significant 'public response', has been claimed to be negligible (Webb 2010). As a result, sociologists, economists and social psychologists increasingly question the relevance of information-deficit and rational actor models of human behaviour and seek alternatives that reflect a range of theoretical, methodological and epistemological commitments (e.g. Sanne 2002; Hobson

2004; Lertzman 2008; Kasper 2009; Uzzell and Räthzel 2009; Shove 2010a, b, c; Soron 2010). The emergence of a variety of alternatives, tensions between them, the potential for points of convergence, and a consideration of their implications for social science and the sustainability agenda, is the focus of subsequent chapters.

Chapter 3 surveys how environmentalism has moved from a fringe, counter-cultural issue to the avowed concern of public, policymakers and corporations in little over fifty years to the point where we can now recognize the mainstreaming of a 'sustainability agenda'. This is a story in which after many years of near-silence, psychology and the social sciences have come to play an increasingly prominent role. This chapter briefly charts some of the early signs of this academic attention and subsequent influence. It will summarize how recent developments across the social sciences have in various ways attempted to advance understandings of the relationship between human and nonhuman nature, understood as in crisis, that more fully accommodate social, psychological and interpenetrating ecological processes (Adams 2012; Kasper 2009; Randall 2009; Uzzell and Räthzel 2009).

The initial focus of the chapter establishes the key claim in this developing literature—that *social context* has been under-theorized and researched in applying psychology to the human-nonhuman nature relationship, particularly envisaged as environmental 'problems' and behavioural 'solutions' (e.g. Webb 2010). Consensus around the need to incorporate 'the social' has to some extent coalesced around a sociological account of individual behaviour as an outcome of the more complex unit of 'social practices' (e.g. Hargreaves 2011; Shove 2010a, b, c, d; Shove and Spurling 2013; Shove and Walker 2010; Shove et al. 2012; Warde 2005), and a detailed overview of this approach is provided in Chapter 4. As a 'social practice approach' has come to dominate the acceptable intellectual understanding of sustainability, and increasingly, the sustainability policy agenda, critical engagement has begun to emerge. Following a summary of the theory, Chapter 5 offers a detailed critical engagement, centred on three issues: meaning, nature and power. From critique emerges the basis for alternative theories and concepts developed in subsequent chapters.

A preliminary assumption of a number of alternatives is that anthropogenic ecological crisis generates profound, and often unconscious anxiety. Chapter 6 explores existential approaches to mortality and finitude as a potential basis for this anxiety. Ernest Becker's influential account of death denial is the starting point for Terror Management Theory, a unique social psychological perspective that provides a bridge between Becker's work and the conceptualization of human responses to knowledge of anthropogenic ecocide. Chapter 7 extends an analysis of the affective dynamics involved via a brief overview and appraisal of psychoanalytically informed theory and research in this area. This research identifies the defensive psychodynamics identified as the cause of inaction and (apparent) indifference in response to ecological degradation. In Chapter 8 we address a minor flurry of psychosocial research that conceptualizes these mechanisms as social and collectively maintained. Empirical studies and theoretical accounts of denial in this context index the artful and highly organized ways through which we assuage uncomfortable emotions arising from partial awareness of human-induced ecological degradation (e.g. Hoggett 2013), and a related sense of our interdependence on the natural world (Randall 2009). Recent articulations wed psychoanalytic concepts with more sociological accounts, both of the management of emotions and the collective organization of defence mechanisms (Norgaard 2011).

Cultural narratives and framings of ecological degradation and related practice clearly play a part in the social organization of denial as it is understood by Norgaard and others. In Chapter 9 the narrative framing of 'environmental problems' is explored in more detail, and the academic attention that the communication of these problems has increasingly garnered. This chapter explores further interdisciplinary work focusing on how framings—understood variously as narratives, discourses and stocks of knowledge—circulate and supposedly infuse personal and social understandings of the nature of ecological crisis and how to respond (Lakoff 2010; Wilk 2010); alongside work advocating the fundamental importance of narrative in meaningful human life. Again this is a disparate body of work, and the key task in this chapter is to establish connections between psychosocial accounts of collectively organized defence mechanisms short on detail when it comes to the particular discourses centrally involved, and work highlighting dominant discourses less clear on the ways in which they are circulated, legitimized, and challenged.

Chapter 10 engages with the thorny issue of challenging dominant narratives in more detail, exploring work that unsettles familiar epistemological and ontological conceptualizations of the relationship between the human and more-than-human world. In particular, it surveys the potential of trans-species psychology, communication and dialogue, for reconceptualizing the relationship between human and nonhuman nature. Engagement here is sympathetic while critical, not least because as a field of study this area epitomizes the difficulties and ambivalences of articulating necessarily opaque understandings of human-nonhuman nature relationship in a culture where such an articulation is 'ideologically occluded' (Kidner 2001).

Accounts, and of course, experiences of trans-species dialogue, thus provide us with one example of reciprocity between the human and more-than-human world which could be meaningfully and culturally articulated (Dickinson 2009). Such discussion ventures into the 'growing currency' of posthumanist discourse and critical engagements with it (e.g. Haraway 2008; Hird 2010; Pickering 2005; Wolfe 2010). This work explicitly questions the privileging and coherence of human nature as distinct from nonhuman entities and receonceptualizes the way human and nonhuman 'encounter and apprehend the world' (Chiew 2014, p. 52). Calls for a more radical vision of human-nonhuman interdependence demands a revisioning of models of participation that would take us into less familiar territory, and test the boundaries of what we take to be a 'psychosocial' perspective. The final chapter is a brief attempt to look forward. It considers what a research agenda might look like as informed by the modest contribution of this book. To do so, it focuses in on the concept of narrative foreclosure as a starting point, before exploring the potential of participatory research for challenging narrative foreclosure.

Notes

1. See for example the environmentalist film *The Age of Stupid* (2009, Dir: Fanny Armstrong), which begins with a majestic sweep of Earth's history accompanied by a countdown clock from its origins to an apocalyptic not-too-distant anti-future; the website and short film *Welcome* to the Anthropocene (International Geosphere-Biosphere Programme 2013) visually maps humanity's impact on Earth across centuries and decades; the British Met Office plot actual and possible changes in climate temperature using a high-definition animation of a rotating Earth (Met Office 2015 http://www.metoffice.gov.

uk/climate-guide/climate-change/impacts/four-degree-rise/map; see also Overseas Development Institute's projections http://www.odi.org/sites/odi. org.uk/files/odi-assets/publications-opinion-files/8634.pdf).

2. Though there is a problem with this logic if recent international polling is to be believed—for it indicates that 'environmental concerns among citizens around the world have been falling since 2009 and have now reached twenty-year lows' (GlobeScan 2013).

References

Adams, M. (2012). A social engagement: How ecopsychology can benefit from dialogue with the social sciences. *Ecopsychology, 4*(3), 216–222.

Brown, S. D., & Stenner, P. (2009). *Psychology without foundations: History, philosophy and psychosocial theory.* London: Sage.

Chiew, F. (2014). Posthuman ethics with Cary Wolfe and Karen Barad: Animal compassion as trans-species entanglement. *Theory, Culture & Society, 31*(4), 51–69.

Dickinson, J. L. (2009). The people paradox: Self-esteem striving, immortality ideologies, and human response to climate change. *Ecology and Society, 14*(1), 34.

Globescan. (2013). Environmental concerns "at record lows": Global poll. http://www.globescan.com/news-and-analysis/press-releases/press-releases-2013/98-press-releases-2013/261-environmental-concerns-at-record-lows-global-poll.html. Accessed 18 Dec 2015.

Gupta, V. (2011). Death and the human condition. In P. Kingsnorth & D. Hine (Eds.), *Dark mountain: Issue 2* (pp. 76–85). Ulverston: Dark Mountain Project.

Haraway, D. J. (2008). *When species meet.* Minnesota: University of Minnesota Press.

Hargreaves, T. (2011). Practice-ing behaviour change: Applying social practice theory to pro-environmental behaviour change. *Journal of Consumer Culture, 11*(1), 79–99.

Hird, M. J. (2010). Indifferent globality: Gaia, symbiosis and 'other worldliness'. *Theory, Culture and Society, 27*(2–3), 54–72.

Hobson, K. (2004). Sustainable consumption in the United Kingdom: The "responsible" consumer and government at "arm's length". *The Journal of Environment & Development, 13*(2), 121–139.

Hoggett, P. (2013). Climate change in a perverse culture. In S. Weintrobe (Ed.), *Engaging with climate change: Psychoanalytic and interdisciplinary perspectives* (pp. 56–71). London: Routledge.

Holloway, J. (2010). *Crack capitalism*. London: Pluto Press.

Kasper, D. V. S. (2009). Ecological habitus: Toward a better understanding of socioecological relations. *Organization and Environment, 22*(3), 311–326.

Kidner, D. W. (2001). *Nature and psyche: Radical environmentalism and the politics of subjectivity*. New York: SUNY Press.

Lakoff, G. (2010). Why it matters how we frame the environment. *Environmental Communication, 4*(1), 70–81.

Lertzman, R. (2008, June 9). The myth of apathy. *The Ecologist*. http://www. theecologist.org/blogs_and_comments/commentators/other_comments/ 269433/the_myth_of_apathy.html. Accessed 18 Dec 2015.

Mazur, L. (2011). Inspiring action: The role of psychology in environmental campaigning and activism. *Ecopsychology, 3*(2), 139–148.

Met Office. (2015). *What is climate change?* London: UK Government. http:// www.metoffice.gov.uk/climate-guide/climate-change. Accessed 18 Dec 2015.

Monroe, M. C. (2003). Two avenues for encouraging conservation behaviors. *Human Ecology Review, 10*, 113–125.

Moser, S. C., & Dilling, L. (Eds.). (2006). *Creating a climate for change*. Cambridge: Cambridge University Press.

Norgaard, K. M. (2006). "We don't really want to know": Environmental justice and socially organized denial of global warming in Norway. *Organization and Environment, 19*(3), 347–370.

Norgaard, K. M. (2011). *Living in denial: Climate change, emotions and everyday life*. Cambridge: MIT Press.

Pickering, A. (2005). Asian eels and global warming: A posthumanist perspective on society and the environment. *Ethics & the Environment, 10*(2), 29–43.

Randall, R. (2009). Loss and climate change: The cost of parallel narratives. *Ecopsychology, 3*, 118–129.

Sanne, C. (2002). Willing consumers—Or locked-in? Policies for a sustainable consumption. *Ecological Economics, 42*(1), 273–287.

Shove, E. (2010a). Sociology in a changing climate. *Sociological Research Online, 15*(3), 12 http://www.socresonline.org.uk/15/3/12.html

Shove, E. A. (2010b). Beyond the ABC: Climate change policy and theories of social change. *Environment and Planning. A, 42*(6), 1273–1285.

Shove, E. A. (2010c). Social theory and climate change questions often, sometimes and not yet asked. *Theory, Culture and Society, 27*(2–3), 277–288.

Shove, E. A. (2010d). Submission to the House of Lords Science and Technology. Select Committee Call for Evidence on Behaviour Change. Available at: http://www.lancs.ac.uk/staff/shove/transitionsinpractice/papers/ShoveHouseof Lords.pdf

Shove, E. A., & Spurling, E. A. (2013). Sustainable practices: Social theory and climate change. In E. A. Shove & N. Spurling (Eds.), *Sustainable practices: Social theory and climate change* (pp. 1–15). London/New York: Routledge.

Shove, E., & Walker, G. (2010). Governing transitions in the sustainability of everyday life. *Research Policy, 39*(4), 471–476.

Soron, D. (2010). Sustainability, self-identity and the sociology of consumption. *Sustainable Development, 18*(3), 172–181.

Stenner, P. (2014). Psychosocial: Qu'est-ce que c'est? *Journal of Psycho-Social Studies, 8*(1), 205–216.

Urry, J. (2010). Sociology facing climate change. *Sociological Research Online, 15*(3), 1.

Uzzell, D., & Räthzel, N. (2009). Transforming environmental psychology. *Journal of Environmental Psychology, 29*(3), 340–350.

Warde, A. (2005). Consumption and theories of practice. *Journal of Consumer Culture, 5*(2), 131–153.

Warde, A. (2013). Sustainable consumption and behaviour change. *Discover Society, 1*. http://www.discoversociety.org/2013/10/01/sustainable-consumption-and-behaviour-change/. Accessed 18 Dec 2015.

Watson, M. (2012). How theories of practice can inform transition to a decarbonised transport system. *Journal of Transport Geography, 24*, 488–496.

Webb, J. (2012). Climate change and society: The chimera of behaviour change technologies. *Sociology, 46*(1), 109–125.

Wilk, R. (2010). Consumption embedded in culture and language: Implications for finding sustainability. *Sustainability: Science, Practice, & Policy, 6*(2), 38–48.

Wolfe, C. (2010). *What is posthumanism?* Minnesota: University of Minnesota Press.

2

Welcome to the Anthropocene

Introduction

Actual and projected ecological crises have been documented with increasing conviction and sophistication since the mid-twentieth century (e.g. Carson 1962; Meadows et al. 1972). What they reveal is the extent to which anthropogenic ecological destruction, climate change (i.e. change in the average temperature Earth's surface) and the depletion of 'natural resources' have become prominent issues since the late twentieth century; accompanied by assertions of growing urgency from an international community of scientists, NGOs, campaign groups and some governments. Parallel narratives of risk and threat are communicated by news media, documentaries, films and art, even if the resulting messages, taken as a whole, are confusing and conflicting.

In an attempt to capture the profound and unprecedented impact of human influence upon the dynamics of life on Earth, an international consensus of geologists declares that our era should be referred to as the 'Anthropocene' (International Geosphere-Biosphere Programme 2013; Oldfield et al. 2014; Zalasiewicz et al. 2010).[1] The term was initially an attempt to convey a sense of the magnitude of human impacts in propor-

© The Author(s) 2016

M. Adams, *Ecological Crisis, Sustainability and the Psychosocial Subject*, DOI 10.1057/978-1-137-35160-9_2

tion to the 'deep time' of Earth's history. Here is how antrhopocene.info, a collaborative project involving researchers and communicators from a number of leading scientific research institutions on global sustainability, define and frame the Anthropocene:

> Our species' whole recorded history has taken place in the geological period called the Holocene—the brief interval stretching back 10,000 years. But our collective actions have brought us into uncharted territory. A growing number of scientists think we've entered a new geological epoch that needs a new name—the Anthropocene. Probably the best-known aspect of our newfound influence is what we're doing to the climate. Atmospheric carbon dioxide may be at its highest level in 15 million years. But this is just one part of the story; we're changing the planet in countless ways. Nutrients from fertilizer wash off fields and down rivers, creating stretches of sea where nothing grows except vast algal blooms; deforestation means vast quantities of soil are being eroded and swept away. Rich grasslands are turning to desert; ancient ice formations are melting away; species everywhere are vanishing. These developments are all connected, and there's a risk of an irreversible cascade of changes leading us into a future that's profoundly different from anything we've faced before. Little by little, we're creating a hotter, stormier and less diverse planet. The Anthropocene is a decisive break from what came before.

In other words, the collective and accumulative outcomes of the activities of (some) human beings is now seriously affecting the nature of life on Earth. In fact, a post-1950 'Great Acceleration' in human impacts is considered by many to be the most suitable starting point for the Anthropocene (Steffen et al. 2015; Hibbard et al. 2006). That influence is understood to pose a profound threat to most of the forms of life we readily identify with.[2]

Interdependence

Donna Haraway refers to Gaia as one of the 'proper icons' of the Anthropocene (we will come to the other one a little later). Gaia stems from James Lovelock's 'daring and provocative' insight that 'the Earth is not just teeming with life. The Earth, in some sense, *is* life. Earth is an organism!' (Ruse 2013a, p. 2; 2013b). Lovelock's hypothesis, later developed in collaboration with Lynn Margulis (e.g. Lovelock and Margulis 1974), was that the Earth is an ongoing self-maintaining system; a complex living process of autopoiesis (Lovelock 2000): 'It suggests that life has conspired in the regulation of the global environment so as to keep conditions comfortable' (Tyrell 2013, p. 3).

The components of the Gaia hypothesis considered the least controversial relate to the *interdependence* of biological, physical and chemical environments, and the coevolution of biota and environment (Kirchner 1989).[3] Scientific studies of specific human practices largely support the ecological premise that anthropogenic effects are radically interdependent, often in ways we are only beginning to understand. To take species extinction as an example, the outcomes of climate change profoundly affect interrelatedness at the level of *interaction between species*. Many ecologists and environmental biologists consider this interaction as vital for future survival:

> Species interact with each other in ways that deeply affect their viability. Certain species impart particularly strong effects on others. Consequently, climate change impacts on these species could initiate cascading effects on other species. In effect, these species act as 'biotic multipliers' of climate change. (Zarnetske et al. 2012, p. 1516)

Zarnetske et al. go on to provide the general example of climate change reducing top predators (in contexts where they are more sensitive to climate variation). This reduction increases the number of herbivores over time, which consequently decreases the number of plants. The result for the 'ecological community' is 'an overall decrease in both species diversity and stability', an example of how 'effects can ripple through an entire food web, multiplying extinction risks along the way' (2012, p. 1517).

'Biotic multiplications' and 'cascading effects' of climate change are evocative additions to both popular and scientific notions of 'tipping points', 'positive feedback', 'climate forcing' and 'runaway climate change' which all speak to the ways in which the profound interrelatedness of anthropogenic ecological effects can exacerbate those effects (Archer and Buffett 2005; Brown 2006; Cox et al. 2000; Doney and Schimel 2007;

Sample 2005). As the Intergovernmental Panel on Climate Change (described in more detail below; subsequently referred to as the IPCC), onerously states, 'the risks of abrupt or irreversible changes increase as the magnitude of the warming increases' (IPCC 2014b, p. 16). In other words, already existing ecological interdependence underpins the multiplication and amplification of anthropogenic effects. Gaia frames this interdependence as planetary self-regulation—a 'balance' that the Anthropocene profoundly upsets.

Scientific Observation and Modelling

Haraway's other icon of the Anthropocene is 'fossil-burning man... extracting and burning fossil fuels as fast as possible' (Haraway 2014). He (the gendered pronoun is intentional) is a suitable symbol of the industrial era and fossil economy that emerged in the late eighteenth century—the starting point of a steep rise in atmospheric concentration of CO_2 and associated global warming (Malm 2015), and a start date for the Anthropocene for some.[4] Furthermore, the burning of fossil fuels (coal, oil, natural gas) is considered the main contribution to the release of CO_2 and other anthropogenic greenhouse gases (e.g. methane, nitrous oxide), so called because they trap heat in the atmosphere, subsequently warming the climate. It is a practice that is integral to most of the activities significantly contributing to anthropogenic climate change: energy production and manufacturing (largely via boiler and furnace emissions), heating, transportation, agriculture, deforestation and waste (IPCC 2014b); and the amplification of many of the above wherever we find economic, population and urban growth (IPCC 2014d). While extracting, refining, transporting and burning fossil fuels is the primary activity here, they unevenly enable a multitude of secondary human activities around work, leisure and the home, such as the consumption of manufactured goods, water and food, turning on a TV or browsing the internet.

Our understanding of the significance of fossil-burning man, and of conveying the ecological interdependence of the Anthropocene more generally, has largely relied on scientific observations and computer models of *climate change*. That 'climate change' should become the predominant imaginary is an interesting point in itself. To some extent it has emerged as a coherent indicator of anthropogenic ecological impacts *because* it has been successfully accommodated within the framework of the physical sciences. The role of science in constructing and framing acceptable narratives and metaphors of ecological crisis is an important issue to which we will return later. First, however, a basic understanding of that science is attempted.[5]

The most recognizable representation of scientific endeavour in this area is the aforementioned Intergovernmental Panel on Climate Change (IPCC). The Panel's membership is made up of 135 governments. It does not conduct climate change research. Its task is the ongoing collation and analysis of the thousands of published scientific papers related to climate change published each year (over 1000 reviewers assessed 30,000 scientific papers for the fifth report, published in 2014). Reviewers are volunteers—scientists and officials from research communities across the globe. The research is derived from historical data, observed changes and predictive models. Key areas of review are divided amongst three Working Groups: the physical science basis of climate change (WGI); negative and positive consequences of climate change, focusing on the vulnerability and adaptability of natural and socio-economic systems, and the interrelationship between these impacts (WGII); and options for climate change mitigation in terms of limiting greenhouse gas emissions and/or removing them from the atmosphere (WGIII, see IPCC 2014a).

The IPCC aims to provide a comprehensive overview of what is known and what is uncertain about the effects and risks associated with climate change; to highlight areas of consensus and disagreement, and to indicate areas where further research is needed (IPCC 2013). The regular IPCC reports are attempts to offer a snapshot of current scientific understanding of the nature and extent of climate change and its impacts. The publication of the latest IPCC Report (AR5) in November 2014 – 'the most comprehensive assessment of climate change ever undertaken' (IPCC 2014b) – provided an update on many of the predictions and observations relating to the issue of climate change made in previous reports made by the IPCC and others stretching back to the 1970s (e.g. Meadows et al. 1972). AR5 is a made up of a number of long and complex documents, but the Synthesis Report (IPCC 2014a) and the Summary for

Policy Makers (IPCC 2014b) are relatively accessible, with degrees of certainty clearly explained. There are also many secondary summaries available (e.g. Friends of the Earth 2015; Met Office 2015; Baraka 2014). On closer inspection, for a scientific community that is 'conservative by nature' (McKibben 2014), there is some unusually strong and committed language in the report.

In AR5 the IPCC reiterate that the atmospheric concentration of CO_2 closely parallels the Earth's surface temperature and that it is because, basically stated, increased levels of CO_2 and other greenhouse gases in the atmosphere make it warmer—they trap heat in (IPCC 2014b). The IPCC consensus accurately reflects the broader scientific consensus—over 97 % of the international scientific community considers the emission of CO_2 and other greenhouse gases into the atmosphere, to be the primary cause of anthropogenic climate change (Cook et al. 2013). CO_2 measurements are therefore considered as suitably reliable for understanding historic, present and future climate change in the context of the Anthropocene.

Greenhouse Gases

Carbon is the basis for all organic life. It has the ability to bond with other elements, to be flexible and yet maintain stability, qualities which make it conducive to molecular being (Archer 2007). It is a central element of all life forms including our own, and to the perpetuation of the conditions that make planetary life possible (Redfern 2003). Billions of tonnes of carbon are constantly being cyclically emitted and absorbed via land, air and water. The balance of the carbon cycle, i.e., between emission and absorption, is central to life on Earth. It is present in different forms—soil, plants, living creatures, dissolved in oceans and as carbon dioxide in the air. The measurement of historic levels of CO_2 in the atmosphere (as parts per million by volume—ppm) is made possible by closely observing marine sediment layers, tree rings, fossilized plants, and most accurately, air bubbles trapped in 'ice cores'—samples taken from ice sheets—in the Arctic and other areas with high mountain glaciers (see Amos 2006; Walker 2004; Weart 2008 for accessible accounts of the process). Ice sheets are formed incrementally from annual buildups of snow, and retain within them observable and detailed information about the climate such as atmospheric gas bubbles. The deeper the ice core, the deeper the historical record—and some samples extracted from a depth of over 3200 m can provide information going back 800,000 years (Parrenin et al. 2007).

Ice core samples and other measurements suggest that the carbon cycle, as reflected in the atmospheric concentration of CO_2, has been relatively stable for the last 400,000 years, varying between 200 and 280 ppm (parts per million), even more so for the last 10,000 years, around 280 ppm (IPCC 2007). In the last 200 years, this figure has rapidly increased from 270 ppm to present day current estimates, which at the time of writing, suggest the level of CO_2 in the atmosphere is 404.83 ppm (see http://co2now. org for regular updates). This increase has seen a particularly sharp rise, since the 1950s, from approximately 310 ppm. For the past decade, there has been an average yearly rise of just over 2.1 ppm, 1.9 ppm the decade before, further reflecting the acceleration of concentration of CO_2 in the atmosphere noted by the IPCC and others (CO_2 now, 2015). These figures mean that 'about half of cumulative anthropogenic CO_2 emissions between 1750 and 2010 have occurred in the last 40 years' (IPCC 2014d, p. 7); with 'larger absolute decadal increases toward the end of this period' (p. 6). Haraway's emphasis on 'fossil-burning man' has a basis in climate science: fossil fuel combustion and industrial process CO_2 emissions are responsible for approximately 78 % of the total increase in greenhouse gases from 1970 to 2010 (IPCC 2014d, p. 6). Current levels are claimed to exceed anything observed in the 800,000-year ice core records (IPCC 2007).

CO_2 is not the only anthropogenic greenhouse gas contributing to the warming of the planet. Methane, nitrous oxide, and some human-made (fluorinated) gases also contribute (IPCC 2007). Shorter life cycles, lower concentrations or other difficulties in detection and measurement make these less reliable indicators of global warming, but there is no doubt that they also contribute.[6] The concentration of non-CO_2 greenhouse gases has also risen steadily over the last 200 years, and at an increasing rate (IPCC 2007). Although less concentrated in the atmosphere, non-CO_2 greenhouse gases are much more effective at trapping heat, so make a significant contribution to the gases responsible for anthropogenic climate change, estimated at approximately 28 % (IPCC 2007, Table 2.1).[7]

Impacts

AR5 reiterates that greenhouse gases are at the highest concentration in recordable history and continue to rise; that human influence on climate is 'clear', that human activity contributes to warming is 'unequivocal', and that the rate of warming is 'unprecedented' and accelerating. That human activity is the main contributor to warming is 'extremely likely' (>95 % probability). Observed changes include a warmer atmosphere, warmer and acidified oceans, reduced amounts of snow and ice (i.e. shrinking Greenland and Arctic ice sheets which together hold the majority of the world's fresh water, reduction in near-surface permafrost, retreating glaciers, declining Arctic sea ice), rising sea levels, and increased precipitation. There are numerous related observations derived from these claims, such as an observable global-scale decrease in number of cold days and nights and increase in warm days and nights; and confident predictions ('virtually certain', 'very likely') about increases in the frequency and duration of extreme temperatures and weather events (e.g. heat waves, droughts, floods, cyclones and wildfires) (IPCC 2014b, p. 11).

Changes such as these are of course closely linked to observed impacts on countless forms of *life*: 'Many terrestrial, freshwater, and marine species have shifted their geographic ranges, seasonal activities, migration patterns, abundances, and species interactions in response to ongoing climate change' (IPCC 2014b, p. 6). This might sound rather benign but of course 'shifts in abundance' often refers to severe threats to life; and to the decimation and destruction of species and their habitat. In fact, some scholars depict the Anthropocene as an age of mass extinction that is already underway, so rapid is the 'shift in abundances' on land, air and sea (Magurran and Dornelas 2010; IPCC 2007). From a standpoint of interrelatedness, extinction as an outcome is intimately bound up in other impacts: migration forces species into new areas of threat, for example, (Kolbert 2014); species interaction, as we have already acknowledged, can impact significantly on the number and balance of species in ecological communities over time (Zarnetske et al. 2012). Hence the IPCC's synopsis: 'a large fraction of species faces increased extinction risk due to climate change during and beyond the 21st century, especially as climate change interacts with other stressors' (IPCC 2014b, p. 13).

Human activity is centrally implicated in causing climate change, but human activity is also significantly affected by it—the Anthropocene rebounds and reverberates back through human life too. There are pervasive issues involved in uncritically homogenizing the role of 'humanity' in ecological crises, be it in terms of responsibility or vulnerability. Recent IPCC documentation centrally acknowledge this unevenness, reflecting the growing research and campaigning focusing on the imbalances and injustices of climate impacts. When it comes to the human impacts of anthropogenic climate change to date, it is clear, to paraphrase the quote that opened this book, that some of us are sitting comfortably while others most definitely are not. Predicted *future* impacts are also dependent to some extent on current and future attempts at mitigation. This brings a further necessary dimension of openness and uncertainty into predictions about future impacts. The IPCC Working Group II's summary of future impacts is largely in terms of risks (2014c, p. 13), predicted to emerge in the next few decades, the rest of the twenty-first century, and beyond with varying degrees of confidence.

The following summary is salutary, and worth reproducing in full, as it is a list of 'high confidence' predictions:

1. Risk of death, injury, ill-health or disrupted livelihoods in low-lying coastal zones and small island developing states and other small islands, due to storm surges, coastal flooding and sea level rise
2. Risks of severe ill-health and disrupted livelihoods for large urban populations due to inland flooding in some regions.
3. Systemic risks due to extreme weather events leading to the breakdown of infrastructure networks and critical services such as electricity, water supply, and health and emergency services.
4. Risk of mortality and morbidity during periods of extreme heat, particularly for vulnerable urban populations and those working outdoors in urban or rural areas.
5. Risk of food insecurity and the breakdown of food systems linked to warming, drought, flooding, and precipitation variability and extremes, particularly for poorer populations in urban and rural settings.

6. Risk of loss of rural livelihoods and income due to insufficient access to drinking and irrigation water and reduced agricultural productivity, particularly for farmers and pastoralists with minimal capital in semi-arid regions.

7. Risk of loss of marine and coastal ecosystems, biodiversity, and the ecosystem goods, functions, and services they provide for coastal live-lihoods, especially for fishing communities in the tropics and the Arctic.

8. Risk of loss of terrestrial and inland water ecosystems, biodiversity, and the ecosystem goods, functions, and services they provide for live-lihoods. (2014c, p. 13)

These risks are in line with many other predictions (e.g. European Commission 2011; Fritsche et al. 2012; Fritze et al. 2008; Nordås and Gleditsch 2007). The IPCC acknowledge that they are particularly chal-lenging 'for the least developed countries and vulnerable communities, given their limited ability to cope' (IPCC 2014c, p. 13). The cautious language here acknowledges global imbalance, whilst avoiding explicit recognition of environmental *injustice*: 'the unfair distribution of envi-ronmental benefits, costs, risks, and opportunities' (Thiele 2013, p. 55). Climate change is often represented as a global risk and universal prob-lem, a process of environmental change that will have potentially cata-strophic impacts across the world, and a problem we are all part of and for which we are collectively responsible. 'But climate change is deeply run through with patterns of inequality. Some are more culpable for cli-mate change and others more at risk... many of the less developed coun-tries of the world, who experience 'double injustice' in that they have little responsibility for climate change but face most of the risks' (Walker 2009, p. 3).

The IPCC and others predict that food production, taking food secu-rity as one example, will be negatively affected by climate change over-all, primarily in reducing crop productivity (Friends of the Earth 2015). The IPCC identify threats to fisheries, wheat, rice and maize produc-tion in particular (IPCC 2014b, p. 14). The demand for food from an increasing population, combined with, in particular, other risks such as pollution reduced groundwater resources and renewable surface water resources (Friends of the Earth 2015; IPCC 2014b). As the above list implies, food productivity is not an abstract threat, but one intimately connected to health risks and therefore, a range of additional threats: 'the ultimate manifestation of reduced yields is health impairment: hunger, under-nutrition, child stunting, susceptibility to infectious diseases, impaired adult health and strength, and premature death' (McMichael et al. 2012, p. 2).

More direct, immediate risks in addition to food security include fallout from extreme weather events, such as 'risks from heat stress, storms and extreme precipitation, inland and coastal flooding, landslides, air pollution, drought, water scarcity, sea-level rise, and storm surges' (IPCC 2014b, p. 16). Anthropogenic climate change and air pollution are 'closely coupled' (US Environmental Protection Agency 2011): each can contribute to the other both directly and indirectly (gases that contribute to warming also pollute the air; gases that pollute the air can further contribute to the greenhouse effect). Poor health and premature death linked to respiratory infections, heart disease, strokes and lung cancer stem from specific forms of particulate air pollution. According to the World Health Organization (WHO), 7 million premature deaths were attributable to polluted air in 2012, 'making it the world' s single biggest environmental health risk' (WHO 2014). Again these are unevenly distributed risks, 'amplified for those lacking essential infrastructure and services or living in exposed areas' (IPCC 2014c, p. 16); 'people who are socially, economically, culturally, politically, institutionally or otherwise marginalized are often highly vulnerable to climate change' (IPCC 2014c, p. 6).

Further, unevenly distributed 'deferred and diffuse' risks include community resilience and mental health effects of extreme weather events, insecure access to food and water and increased displacement of people; the growing likelihood of violent conflicts that follows poverty and economic 'shocks' (McMichael et al. 2008; IPCC 2014c; Burke et al. 2009; Hsiang et al. 2011; United Nations World Food Programme 2014). The interdependence of risks exacerbates injustices: 'Climate change thus acts as a force multiplier, amplifying the negative health impacts of other environmental stressors (such as land degradation, soil nitrification, depletion of freshwater stocks, ocean acidification, and biodiversity loss)' (McMichael et al. 2012, p. 3).

Of course, other species are also subjected to the double injustice of being on the frontline of the effects of anthropogenic ecological degradation while bearing no responsibility for them. Nowhere is this clearer than in the ongoing reality of extinction, suffering and decimation of other species as a direct or indirect result of human activity (e.g. Pearce-Higgins and Green 2014). Kolstad et al. (2014, p. 220) claim that 'if climate change leads to the loss of environmental diversity, the extinction of plant and animal species, and the suffering of animal populations, then it will cause great harms beyond those it does to human beings'. For many the future tense is a misnomer here, it already has.

Mitigation: Where Do We Start?

Mitigation strategies stem from a deceptively simple requirement: reduce CO_2 and other greenhouse gas emissions. This objective is made explicit in Article 2 of the United Nations Framework Convention on Climate Change (cited in IPCC 2014d, p. 4):

> The ultimate objective of this Convention… is to achieve… stabilization of greenhouse gas concentrations in the atmosphere at a level that would prevent dangerous anthropogenic interference with the climate system. Such a level should be achieved within a time frame sufficient to allow ecosystems to adapt naturally to climate change, to ensure that food production is not threatened and to enable economic development to proceed in a sustainable manner.

What the IPCC's most recent report signifies for governments, policymakers and commentators is that we find ourselves at (another) decisive moment, but there is now also the unavoidable implication that we have left it too late. None of the paths we can take moving forward can avoid dealing with the impacts of climate change, for the planet has already warmed, and will continue to do so. Instead, the scenarios represent how various attempts at mitigation and adaptation might interrelate with existing risks, vulnerabilities and resilience. The agreed target here, to reiterate, is the reduction of CO_2 emissions and other greenhouse gases, in order to restrict the temperature increase from pre-industrial levels to below 1.5°C and therefore, to have a chance of adapting to, and eventually containing within liveable limits, inevitable global warming and its impacts. As Squarzoni states it across his elegant panels (2014, p. 188):

> We stand at a crossroads. We cannot stop the planet heating up over the next few decades. But the scale of the impending disruptions depends on how we react. To stabilize the amount of CO_2 in the atmosphere, we have to bring our emissions back down to a level where they can be absorbed by nature—oceans, plant life, the soil. To do that... we have to reduce our emissions by 75 % [estimates vary between 40-70 % according to the more cautious IPCC 2014d, p. 10] before 2050. Where do we start?

So, where *do* we start? IPCC suggestions for mitigation follow a simple enough logic: make changes in the sectors emitting the most greenhouse gases. However, these sectors overlap with a complex set of 'drivers' that include, according to IPCC, 'population size, economic activity, lifestyle, energy use, land-use patterns, technology and climate policy' (IPCC 2014b).[8] The IPCC index various potential interventions targeting these drivers: 'large-scale' changes in energy systems (e.g. improved efficiency, increase supply from renewables and carbon dioxide capture storage technology, changes in consumption patterns); land use (e.g. afforestation, management of crops, livestock and soil); transport mode (changes in transport mode, energy efficiency and vehicle performance improvements); buildings (use of new technologies, changes in lifestyle, culture and behaviour towards lower energy demand); industry (energy efficiency, improvements in production, re-use and recycling of materials); and waste (treatment technologies; recovering of energy from waste; efficient recycling).

Some of these changes are already underway and have achieved reductions in CO_2 emissions (IPCC 2014b). However, the nature and extent of mitigation attempts required to be 'successful' (i.e. have a hope of keeping warming below 1.5°C) over the remainder of this century is complex, contested and uncertain—unsurprising considering the magnitude of change required. The IPCC address this uncertainty and complexity by presenting four scenarios. These are referred to collectively as the rather unwieldy 'Representative Concentration Pathways' (RCPs) (IPCC 2014a; see Van Vuuren et al. 2011; Wayne 2013 for an overview). The projections are complex, but each pathway reflects different degrees of global warming and CO_2 atmospheric concentration leading up to 2100 from a pre-industrial baseline, depending on the extent of mitigation. Temperatures vary between 2.6 and 8.5°C, broadly representing 'stringent mitigation scenarios' at the lower end, and a 'baseline scenario' at the other, in which there are no additional efforts to constrain emissions beyond what are presently in place. In between these are 'intermediate scenarios' reflecting varied attempts at mitigation and allowing for uncertainties in the calculations. The IPCC scenarios reflect the extent to which they are implemented, and at what timescales (2014d, p. 12).

Notwithstanding the IPCC's attempt to calculate future scenarios, there is a great deal of uncertainty over what changes are needed and to what extent they are achievable. According to many commentators, for example, 'stringent mitigation scenarios' involve leaving the overwhelming majority of the Earth's remaining fossil fuel reserves in the ground and adopting alternative energy production immediately, although this is not explicitly articulated by the IPCC (Monbiot 2015). The details of the scenarios suggest that we may have already moved beyond a point where it is possible to limit warming to 1.5°C, though there is the hope that following a 'stringent mitigation scenario' (RP2.6) CO_2, and therefore climate, could be stabilized at a 'safe' level by 2100, *after* an initial overshoot (IPCC 2014a). The possibility that it is too late to stabilize warming at 1.5°C by reducing emissions has led to a revival of hope in some quarters that technological 'geo-engineering' solutions can be utilized to do the job for us, or manage climate more directly (for overviews of arguments for and against see Corner and Pidgeon 2010; Gardiner 2010; Preston 2013). None of these appear to be genuine solutions in light of the necessary timescales and complexities of climate change representations as they are represented by the IPCC (Hulme 2014).

The complexity of addressing anthropogenic climate change—of agreeing what changes are needed and how they might be achieved—is exacerbated by two further factors. First, the way scientific understandings frame anthropogenic ecological crisis in general and climate change in particular as a problem in the first place is conducive to some ways of engaging individually and collectively with that problem, but a hindrance to others (Erickson 2015). Second is the significance of political, psychological, social and cultural processes involved in generating the necessary changes (or perpetuating the status quo), factors that the IPCC and other scientific bodies increasingly recognize as salient.

Scientising the Anthropocene

The first factor that exacerbates the complexity of deciding what changes are needed and how they might be achieved is the way scientific understandings frame the Anthropocene as a problem in the first place. We rely on scientific frames to know about climate change at all. This reliance stems to some degree from the unobtrusiveness of climate change for many of 'us' in that it does not readily appear to be perturbing everyday life experience (Schäfer and Schlichting 2014). Žižek illustrates this 'obstacle' with the following imagined attitude 'I know very well (that global warming is a threat to the entire humanity), but nonetheless... (I cannot really believe it). It is enough to see the natural world to which my mind is connected: green grass and trees, the sighing of the breeze, the rising of the sun... can one really imagine that this will be disturbed?' (Žižek 2011, p. 445–6). Our dependence on scientific knowledge is also a consequence of the inherent complexity and 'incalculableness' of the methods and data involved (Beck 2013; Schäfer and Schlichting 2014; Weart 2008, 2010); and the cultural authority of scientific narratives (Erickson 2015).

This does not mean that when people *do* directly experience the impacts of climate change, they do or will experience 'it' or conceptualize 'it' as climate change. How do we understand climate change if we do not 'know' it is 'climate change'?

It is only because of formal science that we know about climate change. Without this scientific knowledge we would experience weather, and, probably, more and more extreme weather events as global warming proceeds apace. At a certain point, unchecked global warming could render large parts of the planet uninhabitable either through sea-level rises or desertification. Without formal science we would be unaware of why this was happening, although we would still experience the phenomena. (Erickson 2015, p. 278)

So science provides us with a narrative or at least a series of interconnected metaphors – greenhouse gases, global warming and climate change – with which we are invited to hold together the otherwise incoherent and fragmented experience of weather, landscape, other species, with scientific evidence and its various mediations. Scientific knowledge understood in this way is providing an invaluable service. Without it, it is highly unlikely that 'we' would 'notice' patterns at all, let alone frame them in an overarching narrative. Whatever the wider social and political response, the Intergovernmental Panel on Climate Change and other bodies are undoubtedly utilizing science to articulate as clearly as possible the extent of climate change, and its current and potential impacts, the benefits of mitigations and the dangers of inertia, and attached injustices and imbalances. However, any narrative is inevitably reductive: it both excludes possible understandings and courses of action and facilitates others:

> It is remarkable that we keep thinking of problems that are caused by humans, that inflict harm on humans (and the life support systems on which they depend), and that can only be solved by humans, in terms of their biophysical nature—as matters of molecules, shifts in atmospheric dynamics or ecosystem interactions, imbalances in elemental cycles, or merely as collapsing environmental systems. (Hackmann et al. 2014, p. 654)

The danger in scientising the Anthropocene lies in the reframing of a social problem as a physical one. It has led to a focus on physical climate change in particular—that the issue is the global climate, that the focal point is the mean temperature of the Earth's surface, and that the most significant event is the rise in that temperature (Erickson 2015). Consequently, if we could reduce this temperature, then we would fix the problem. This is 'the official position of planet Earth… we can't raise the temperature more than 2°C [1.5°C since the 2015 United Nations Climate Change Conference in Paris]—it's become the bottomest of bottom lines' (McKibben 2012). This narrative frames the problem *and the possible solutions* in particular ways (Erickson 2015, p. 280): stopping warming. The UN-mandated means to this end to date has been reducing greenhouse gas emissions as described above. The possible failure of this solution has led to the revival of interest in geoengineering solutions in some quarters, described above, whereby the climate is controlled via human technologies (Klein 2014). The point here is that scientific frameworks encourage us to look to science 'to fix what is, in essence, a social and economic problem' (Erickson 2015, p. 283).

Another problem relating to our reliance on climate science is that there is no straightforward fit between scientific ways of knowing and the everyday forms of individual and social knowing and acting. The IPCC recognize that scientific knowledge cannot simply be grafted on to the ways and means through which we understand the world around us, and our relationship to it. This disjunction contributes to many a slippage:

> People's perceptions and understanding of climate change do not necessarily correspond to scientific knowledge... because they are more vulnerable to emotions, values, views, and (unreliable) sources... People are likely to be misled if they apply their conventional modes of understanding to climate change. (Kolstad et al. 2014, p. 254)

This quote highlights an unfortunate tendency in climate science communication to promote a hierarchical binary of objective, deliberative and pristine 'scientific knowledge' on the one hand, and the intuitive, misleading and messy nature of 'conventional' knowledge on the other. However, the problem is not merely how we effectively transport knowledge from one domain ('scientific knowledge') to another ('everyday understandings'), mindful of the obstacles of 'conventional understanding'. Acknowledging 'vulnerabilities' might be an advance on the information-deficit model, which assumes that filling up individual decision makers with the right knowledge will lead to the right attitudes, followed by the right behaviours. It acknowledges the 'gap' between attitude and behaviour, as constantly filled out by a proclivity for unreliable and misleading emotions, values and attitudes. Solutions are therefore, derived from behavioural economics, encouraging the manipulation of information and activities to bypass or overcome these proclivities, so that people can, at last, see what science is telling us. Amongst other problems, such an approach overestimates the ability of scientific knowledge to *matter* to people, in ways that might transform individual and collective behaviour. It is in this sense that Candy Callison speaks of climate change as a 'double bind'. This is a helpful conceptualization, worth citing at length:

> The first half of the bind is this: climate change is ultimately an amalgam of scientific facts based on modelling, projections and empirical observations of current and historical records found in tree rings, coral reefs, ice cores, sea ice cover and other forms of data. Acceptance of the premise of climate change requires a fidelity to and trust in scientific methods, as well as institutional processes like the IPCC that collate, evaluate, and summarise global research related to climate change... The second half of the double bind is that in order to engage diverse publics and discuss ramifications and potential actions... climate change must become much more than an IPCC-approved fact *and* maintain fidelity to it at the same time. It must promiscuously inhabit the spaces of ethics, morality and other community-specific rationales for actions whilst resting on scientific methodology and institutions that prize objectivity and detachment from politics, religion and culture. (Callison 2014, p. 14)

Understanding that climate change comes to matter to people through engagement with other knowledge frameworks, and cultural and collective endeavours is a psychosocial perception. To inhabit the spaces of 'ethics, morality and other community-specific rationales for actions', scientific knowledge must be protean, but also subjective, impassioned and attached. Exploring this terrain, as we will in this book, means seeking to understand how a capacity to engage is refracted through embodied experience, relationships, cultural conventions and material arrangements.

Addressing the Psychosocial

The second factor noted above was the role of political, psychological, social and cultural processes in generating the necessary changes (or perpetuating the status quo). The reach of the Anthropocene stretches well beyond the outcomes of a warming climate to a web of interrelated effects. Power generation, industry, transportation, heating and cooling homes, businesses and shops, deforestation, waste practices, agriculture, hunting and poaching, land and building usage create ecological problems that interrelate with climate change but create problems in their own right—pollution of air, water and land, and harm to many species and their habitats, including human beings, which is often irreversible. The scale of these effects is enormous, warranting the epochal definition. The Anthropocene is a profoundly unsettling narrative that places us firmly on terra firma yet takes us well beyond the physical sciences. It can no longer be considered a niche concern of environmentalists or 'greens'. It cuts through countless social science topics such as identity, power, globalization, violence, justice and equality, consumerism, family and relationships, and has social and personal corollaries that range from 'our everyday choices about resource use and lifestyles, through how we adjust to an unprecedented rate of environmental change, to our role in debating and enacting accompanying social transitions' (Whitmarsh et al. 2013, p. 7).

In this way, our everyday activities are implicated in climate change and vice versa; as Squarzoni's remarkable illustrated account of his own journey through climate change information makes clear:

> We live in a world of fictions. A fable disconnected from reality. The material prosperity we've enjoyed [some of us] over the last two centuries has been dependent on cheap and abundant energy... the accumulation of consumer goods... and the destruction of nature. Whether we like it or not our way of life and CO_2 emissions are inextricably linked. Whether we like it or not... there are greenhouse gas emissions in every part of our life, from our food, our cars, our homes, our pastimes. All our activities are part of the climate crisis, all our wants, every product we purchase, the way we eat, get around, keep warm. Eradicating so much CO_2 from our way of life won't be easy. What do we cut out first? (Squarzoni 2014, pp. 215–218)

If the inextricable interconnections between livelihoods, lifestyles, behaviour and culture matter then climate change is not simply a scientific issue. It is an economic, societal, political problem of the highest order, demanding levels of change in which the psychological and the social are mutually implicated. These factors, combined, hint at the potential value of a psychosocial perspective for understanding the scale of change required to begin to address the realities of anthropogenic ecological degradation. Envisaging change, demanding and enacting it as well as disavowing and resisting it, involves 'the psychological'. It evokes our fears, hopes, creative capacities, and prejudices; it depends on and configures relationships, conventions and habitual ways of 'going on'. These processes animate social and cultural life, but the psychological is also the embodiment of the social, cultural, political, historical and material. The psychological is preceded by the ways ecological crisis is discursively framed in relation to existing narratives, conventions and technologies.

The psychosocial approach adhered to in this book attempts to understand such dynamics as simultaneously psychological *and* sociocultural, to the point that the distinction becomes purposefully uncertain. The ways climate change is communicated and made sense of; lived experience of the impacts of ecological degradation; anxiety over ways of life we might be emotionally invested in but perceive to be under threat—these are all issues relevant to the scale of change anthropogenic ecological degradation demands. All interweave the psychological and the social, individual and collective, intimate and public, in ways which question the binary distinctions on which they are built. Accordingly, much of this book employs concepts and theories that transcend or at least challenge a ready distinction between social and psyche. The next chapter explores the social and cultural context in which knowledge and understanding of anthropogenic ecological degradation have arisen in more detail.

Notes

1. For some commentators this term fails to differentiate between different human eras, societies and classes – 'humanity as an un-differentiated whole' (Moore 2014a, p. 2); and attendant relations of power and capital stretching back to the fifteenth century; hence Moore prefers the term 'capitalocene'. Though see Angus (2014) and Haraway (2014) for further critique.
2. To paraphrase Hird (2010), microbiotic life, which makes up the bulk of living matter, is profoundly indifferent to human activity.
3. The Gaia hypothesis has attracted fluctuating scientific respectability, debate and controversy (Lenton 1988; Tyrell 2013), and is far from established, and

there is not the space here to engage with relevant scientific debates or competing hypotheses in detail (see Kirchner 2002; Ruse 2013a, b; Tyrell 2013 for detailed critical engagements).

4. Though the most suitable start date is still debated. The date of the first nuclear tests is an interesting suggestion, as the fallout from these tests will leave lasting geological records indicating anthropogenic influence (Biello 2015; Zalasiewicz 2009).

5. These are the basic scientific understandings of the material foundations of climate change, poorly conveyed, no doubt, by a non-scientist. The references and further reading suggestions can lead you to a more detailed engagement. For anyone interested, but daunted, I can strongly recommend Philippe Squarzoni's *Climate Changed* (2014)—a remarkable piece of scholarship, an engaging work of art and a great guide for, as the subtitle suggests, a 'personal journey through the science'.

6. For example, methane emissions have doubled as a by-product of the energy industries, agriculture (the latter including, infamously, the fermentation of food in the digestive systems of sheep, cows and other livestock, but also rice growing and forest clearing), and landfill (US Environmental Protection Agency 2010). Nitrous oxide is released by the uses of fertilizer in agriculture, as well as being a by-product of burning fossil fuels and other industrial and agricultural processes. Fluorocarbon gases are not emitted naturally— they are solely the product of human activity, again derived from industrial processes, refrigeration and consumer products (IPCC 2007; US Environmental Protection Agency 2010).

7. The US Government's Environment Protection Agency estimates a rise of 10 % in non-CO_2 greenhouse gases between 1990 and 2005, and a predicted 43 % rise between 2005 and 2030 in a 'business-as-usual scenario', i.e. 'in which currently achieved reductions are incorporated but future mitigation actions are included only if either a regulation, well-established programme, or an international sector agreement is in place' (Environment Protection Agency 2012, p. 10).

8. The IPCC also single out urbanization as a significant threat because it 'locks-in' a range of practices considered central to greenhouse gas emissions and therefore any hopes for mitigation (2014d, p. 23):

> Infrastructure and urban form are strongly interlinked, and lock-in patterns of land use, transport choice, housing, and behaviour. Effective mitigation strategies involve packages of mutually reinforcing policies,

including co-locating high residential with high employment densities, achieving high diversity and integration of land uses, increasing accessibility and investing in public transport and other demand management measures.

However, contemporary urbanization, likely to continue up to and beyond 2050 also presents a 'window of opportunity' (p. 23), as new infrastructure increases the possibilities for a more integrated approach to the changes needed across different sectors.

References

Amos, J. (2006, September 4). Deep ice tells long climate story. *BBC News.* http://news.bbc.co.uk/1/hi/sci/tech/5314592.stm. Accessed 18 Dec 2015.

Angus, I. (2014, September 17). The problem with 'capitalocene'. *Climate and Capitalism blog.* http://climateandcapitalism.com/2014/09/17/problem-capitalocene/. Accessed 18 Dec 2015.

Archer, D. (2007). *Global warming: understanding the forecast.* Oxford: Oxford University press.

Archer, D., & Buffett, B. (2005). Time-dependent response of the global ocean clathrate reservoir to climatic and anthropogenic forcing. *Geochemistry Geophysics Geosystems, 6*(3), Q03002.

Baraka, H. (2014, October 30). *10 important findings from the IPCC Reports.* 350. org. http://350.org/10-important-findings-from-the-ipcc-reports/. Accessed 18 Dec 2015.

Beck, U. (2013). Living and coping with a world risk society. In D. Innerarity & J. Solana (Eds.), *Humanity at risk* (pp. 11–18). London: Bloomsbury.

Biello, D. (2015, February 10). *Nuclear blasts may prove best marker of humanity's geologic record.* Scientific American. http://www.scientificamerican.com/article/nuclear-blasts-may-prove-best-marker-of-humanity-s-geologic-record-in-photos/. Accessed 18 Dec 2015.

Brown, P. (2006, November 18). How close is runaway climate change? *The Guardian.* http://www.theguardian.com/environment/2006/oct/18/bookextracts.books. Accessed 18 Dec 2015.

Burke, M. B., Miguel, E., Satyanath, S., Dykema, J. A., & Lobell, D. B. (2009). Warming increases the risk of civil war in Africa. *Proceedings of the National Academy of Sciences, 106*(49), 20670–20674.

Callison, C. (2014). *How climate change comes to matter. The communal life of facts.* Durham: Duke University Press.

Carson, R. (1962a). *Silent spring.* New York: Houghton Mifflin Company.

Cook, J., Nuccitelli, D., Green, S. S., Richardson, M., Winkler, B., Painting, R., Way, R., Jacobs, P., & Skuce, A. (2013). Quantifying the consensus on anthropogenic global warming in the scientific literature. *Environmental Research Letters* (IOP Publishing), *8*(2).

Corner, A., & Pidgeon, N. (2010). Geoengineering the climate: The social and ethical implications. *Environment: Science and Policy for Sustainable Development, 52,* 24–37.

Cox, P. M., Betts, R. A., Jones, C. D., Spall, S. A., & Totterdell, I. J. (2000). Acceleration of global warming due to carbon-cycle feedbacks in a coupled climate model. *Nature, 408*(6809), 184–187.

Doney, S. C., & Schimel, D. S. (2007). Carbon and climate system coupling on timescales from the Precambrian to the Anthropocene. *Annual Review of Environment and Resources, 32,* 31–63.

Erickson, M. (2015a). *Science, culture and society: Understanding science in the 21st century* (2 ed.). Cambridge: Polity.

Environmental Protection Agency (EPA). (2010). *Methane and nitrous oxide emissions from natural sources.* Washington, DC: Environmental Protection Agency.

European Commission. (2011). The IISS transatlantic dialogue on climate change and security. www.iiss.org/programmes/transatlantic-dialogue-on-climate-change-and-security/. Accessed 18 Dec 2015.

Friends of the Earth. (2015). *Climate change science.* http://www.foe.co.uk/campaigns/climate/climate_change_evidence. Accessed 18 Dec 2015.

Fritsche, I., Cohrs, J. C., Kessler, T., & Bauer, J. (2012). Global warming is breeding social conflict: The subtle impact of climate change threat on authoritarian tendencies. *Journal of Environmental Psychology, 32,* 1–10.

Fritze, J. G., Blashki, G. A., Burke, S., & Wiseman, J. (2008). Hope, despair and transformation: Climate change and the promotion of mental health and wellbeing. *International Journal of Mental Health Systems, 2,* 13.

Gardiner, S. M. (2010). Is "arming the future" with geoengineering really the lesser evil? Some doubts about the ethics of intentionally manipulating the climate system. In S. M. Gardiner (Ed.), *Climate ethics: Essential readings* (pp. 284–314). Oxford: Oxford University Press.

Hackmann, H., Moser, S. C., & St. Clair, A. L. (2014). The social heart of global environmental change. *Nature Climate Change, 4,* 653–655.

Haraway, D. (2014, September 5). Anthropocene, Capitalocene, Chthulucene: Staying with the Trouble. *Antrhopocene: Arts of living with a damaged planet*, AURA: Aarhus University Research on the Anthropocene. https://vimeo.com/97663518

Hibbard, K. A., Crutzen, P. J., Lambin, E. F., et al. (2006). Decadal interactions of humans and the environment. In R. Costanza, L. Graumlich, & W. Steffen (Eds.), *Integrated history and future of people on earth. Dahlem Workshop Report* 96, pp. 341–375.

Hird, M. J. (2010). Indifferent globality: Gaia, symbiosis and 'other worldliness'. *Theory, Culture and Society, 27*(2–3), 54–72.

Hsiang, S. M., Meng, K. C., & Cane, M. A. (2011). Civil conflicts are associated with the global climate. *Nature, 476*, 438–441.

Hulme, M. (2014). *Can science fix climate change? A case against climate engineering*. Cambridge: Polity.

International Geosphere-Biosphere Programme. (2013). Welcome to the anthropocene. http://www.anthropocene.info/en/home. Accessed 15 May 2016.

IPCC. (2007). *Fourth Assessment. Working Group I: The Physical Science Basis*. http://www.ipcc.ch/publicationsanddata/ar4/wg1/en/tssts-2-1-1.html. Accessed 18 Dec 2015.

IPCC. (2013). http://www.ipcc.ch/news_and_events/docs/factsheets/FS_what_ipcc.pdf

IPCC. (2014a). Fifth assessment report (AR5). http://www.ipcc.ch/

IPCC. (2014b). Fifth assessment report (AR5): Summary for policy makers. http://www.ipcc.ch/pdf/assessment-report/ar5/syr/SYRAR5SPMcorr2.pdf

IPCC. (2014c). In C. B. Field, V. R. Barros, D. J. Dokken, K. J. Mach, M. D. Mastrandrea, T. E. Bilir, M. Chatterjee, K. L. Ebi, Y. O. Estrada, R. C. Genova, B. Girma, E. S. Kissel, A. N. Levy, S. MacCracken, P. R. Mastrandrea, & L. L. White (Eds.), *Climate change 2014: Impacts, adaptation, and vulnerability. Part A: Global and sectoral aspects. Contribution of working group II to the fifth assessment report of the intergovernmental panel on climate change* (p. 1132). Cambridge/New York: Cambridge University Press.

IPCC. (2014d). In O. Edenhofer, R. Pichs-Madruga, Y. Sokona, E. Farahani, S. Kadner, K. Seyboth, A. Adler, I. Baum, S. Brunner, P. Eickemeier, B. Kriemann, J. Savolainen, S. Schlömer, C. von Stechow, T. Zwickel, & J. C. Minx (Eds.), *Climate change 2014: Mitigation of climate change. Contribution of working group III to the fifth assessment report of the intergovernmental panel on climate change*. Cambridge/New York: Cambridge University Press.

Kirchner, J. W. (1989). The Gaia hypothesis: Can it be tested? *Review of Geophysics, 27*, 223–235.

Kirchner, J. W. (2002). The Gaia hypothesis: Fact, theory, and wishful thinking. *Climate Change, 52*, 391–408.

Klein, N. (2014). *This changes everything. Capitalism vs. the climate.* Harmondsworth: Penguin.

Kolbert, E. (2014). *The Sixth extinction: An unnatural history.* London: Picador.

Kolstad, C., Urama, K., Broome, J., Bruvoll, A., Cariño Olvera, M., Fullerton, D., Gollier, C., Hanemann, W. M., Hassan, R., Jotzo, F., Khan, M. R., Meyer, L., & Mundaca, L. (2014). Social, economic and ethical concepts and methods. In O. Edenhofer, R. Pichs-Madruga, Y. Sokona, E. Farahani, S. Kadner, K. Seyboth, A. Adler, I. Baum, S. Brunner, P. Eickemeier, B. Kriemann, J. Savolainen, S. Schlömer, C. von Stechow, T. Zwickel, & J. C. Minx (Eds.), *Climate change 2014: Mitigation of climate change. Contribution of Working Group III to the fifth assessment report of the Intergovernmental Panel on Climate Change.* Cambridge, UK/New York: Cambridge University Press.

Lenton, T. M. (1998). Gaia and natural selection. *Nature, 394*(6692), 439–447.

Lovelock, J. (2000) [1979]. *Gaia: A new look at life on earth* (3rd ed.). Oxford: Oxford University Press.

Lovelock, J. E., & Margulis, L. (1974). Atmospheric homeostasis by and for the biosphere: The Gaia hypothesis. *Tellus, 26*, 2.

Magurran, A. E., & Dornelas, M. (2010). Introduction: Biological diversity in a changing world. *Philosophical Transactions B, 365*, 3593–3597.

Malm, A. (2015). *Fossil capital. The rise of steam power and the roots of global warming.* London: Verso.

McKibben, B. (2012, August 2). Global warming's terrifying new math. *Rolling Stone,* Issue 1162. http://www.rollingstone.com/politics/news/global-warmings-terrifying-new-math-20120719#ixzz3SrbRS2SV. Accessed 18 Dec 2015.

McKibben, B. (2014, November 2). The IPCC is stern on climate change – But it still underestimates the situation. *The Guardian.* https://www.theguardian.com/environment/2014/nov/02/ipcc-climate-change-carbon-emissions-underestimates-situation-fossil-fuels. Accessed 10 November 2016.

McMichael, A. J., Friel, S., Nyong, T., & Corvalan, C. (2008). Global environmental change and health: Impacts, inequalities, and the health sector. *British Medical Journal, 336*, 191–194.

McMichael, T., Montgomery, H., & Costello, A. (2012). Health risks, present and future, from global climate change. *BMJ, 344,* e1359.

Meadows, D. H., Meadows, G., Randers, J., & Behrens, W. W. (1972). *The limits to growth.* New York: Universe Books.

Met Office. (2015). *What is climate change?* London: UK Government. http://www.metoffice.gov.uk/climate-guide/climate-change. Accessed 18 Dec 2015.

Monbiot, G. (2015, March 10). Keep fossil fuels in the ground to stop climate change. *The Guardian.* http://www.theguardian.com/environment/2015/mar/10/keep-fossil-fuels-in-the-ground-to-stop-climate-change. Accessed 18 Dec 2015.

Moore, J. W. (2014a). *The Capitalocene: Part I: On the nature and origins of our ecological crisis.* http://www.jasonwmoore.com/uploads/TheCapitalocene PartIJune2014.pdf. Accessed 18 Dec 2015.

Moore, J. W. (2014b). *The Capitalocene part II: Abstract social nature and the limits to capital.* http://www.jasonwmoore.com/uploads/TheCapitalocene PartIIJune2014.pdf. Accessed 18 Dec 2015.

Moore, J. W. (2014c). The end of cheap nature or: How I learned to stop worrying about 'the' environment and love the crisis of capitalism. In C. Suter & C. Chase-Dunn (Eds.), *Structures of the world political economy and the future of global conflict and cooperation* (pp. 285–314). Berlin: LIT.

Nordås, R., & Gleditsch, N. P. (2007). Climate change and conflict. *Political Geography, 26*(6), 627–638.

Oldfield, F., Barnosky, A. D., Dearing, J., Fischer-Kowalski, M., McNeill, J., Steffen, W., & Zalasiewicz, J. (2014). The Anthropocene Review: Its significance, implications and the rationale for a new transdisciplinary journal. *The Anthropocene Review, 1,* 3–7.

Parrenin, F., Barnola, J.-M., Beer, J., Blunier, T., Castellano, E., Chappellaz, J., Dreyfus, G., Fischer, H., Fujita, S., Jouzel, J., Kawamura, K., Lemieux-Dudon, B., Loulergue, L., Masson-Delmotte, V., Narcisi, B., Petit, J.-R., Raisbeck, G., Raynaud, D., Ruth, U., Schwander, J., Severi, M., Spahni, R., Steffensen, J. P., Svensson, A., Udisti, R., Waelbroeck, C., & Wolff, E. (2007). The EDC3 chronology for the EPICA Dome C ice core. *Climate of the Past, 3,* 485–497.

Pearce-Higgins, J. W., & Green, R. E. (2014). *Birds and climate change: A global review of the impact of climate change on birds.* Cambridge: Cambridge University Press.

Preston, C. J. (2013). Ethics and geoengineering: Reviewing the moral issues raised by solar radiation management and carbon dioxide removal. *Wiley Interdisciplinary Reviews: Climate Change, 4,* 23–37.

Redfern, M. (2003). *The Earth: A very short introduction.* Oxford: Oxford University Press.

Ruse, M. (2013a). *The Gaia hypothesis: Science on a pagan planet.* Chicago: University of Chicago Press.

Ruse, M. (2013b, January 14). Earth's holy fool? *Aeon.* http://aeon.co/magazine/science/michael-ruse-james-lovelock-gaia/. Accessed 18 Dec 2015.

Sample, I. (2005, August 11). Warming hits 'tipping point'. *The Guardian.* http://www.theguardian.com/environment/2005/aug/11/science.climatechange1. Accessed 18 Dec 2015.

Schäfer, M. S., & Schlichting, I. (2014). Media representations of climate change: A meta-analysis of the research field. *Environmental Communication, 8*(2), 142–160.

Squarzoni, P. (2014). *Climate changed: A personal journey through the science.* New York: Abrams ComicArt.

Steffen, W., Broadgate, W., Deutsch, L., Gaffney, O., & Ludwig, C. (2015). The trajectory of the Anthropocene: The Great Acceleration. *The Anthropocene Review, 2,* 81–98.

Thiele, L. P. (2013). *Sustainability.* Cambridge: Polity Press.

Tyrell, T. (2013). *On Gaia: A critical investigation of the relationship between life and earth.* Princeton: Princeton University Press.

U.S. Environmental Protection Agency. (2012). *Summary report: Global anthropogenic NonCO2 greenhouse gas emissions: 1990–2030.* Office of Atmospheric Programs Climate Change Division Washington, DC. http://www.epa.gov/climatechange/EPAactivities/economics/nonco2projections.html. Accessed 18 Dec 2015.

United Nations World Food Programme. (2014). *Climate impacts on food security and nutrition: A review of existing knowledge.* United Nations. http://www.wfp.org/content/climate-impacts-food-security-and-nutrition-review-existing-knowledge. Accessed 18 Dec 2015.

US Government EPA. (2011). *Climate change and air pollution.* http://www.epa.gov/airquality/airtrends/2011/report/. Accessed 18 Dec 2015.

Van Vuuren, D. P., Edmonds, J., Kainuma, M., Riahi, K., Thomson, A., Hibbard, K., & Rose, S. K. (2011). The representative concentration pathways: An overview. *Climatic Change, 109,* 5–31.

Walker, G. (2004). Palaeoclimate: Frozen time. *Nature, 429,* 596–597.

Walker, G. (2009). Academic perspectives on climate change and social justice. In *ESRC (2009) Seminar series: Mapping the public policy landscape 'How will climate change affect people in the UK and how can we best develop an equitable response?'* (pp. 3–5). Swindon: Joseph Rowntree Foundation, Local Government Association, ESRC.

Wayne, G. P. (2013). The beginner's guide to representative concentration pathways. *Skeptical Science.* http://www.skepticalscience.com/rcp.php. Accessed 18 Dec 2015.

Weart, S. (2008). *The discovery of global warming* (2nd ed.). Harvard: Harvard University Press.

Weart, S. (2010). *The discovery of global warming: A hyperlinked history of climate change science.* http://www.aip.org/history/climate/summary.htm. Accessed 18 Dec 2015.

Whitmarsh, L., O'Neill, S., & Lorenzoni, I. (2013). Public engagement with climate change: What do we know and where do we go from here? *International Journal of Media & Cultural Politics, 9*(1), 7–25.

World Health Organization. (2014). *Burden of disease from ambient and household air pollution.* http://www.who.int/phe/healthtopics/outdoorair/databases/en/. Accessed 18 Dec 2015.

Zalasiewicz, J. (2009). *The earth after us: What legacy will humans leave in the rocks?* Oxford: Oxford University Press.

Zalasiewicz, J., Williams, M., Steffen, W., & Crutzen, P. (2010). The new world of the Anthropocene. *Environment Science and Technology, 44,* 2228–2231.

Zarnetske, P. L., Skelly, D. K., & Urban, M. C. (2012). Biotic multipliers of climate change. *Science, 336*(6088), 1516–1518.

Žižek, S. (2011). *Living in the end times.* London: Verso.

3

Ecological Crisis Through a Social Lens

Introduction

> The environmental challenges that confront society are unprecedented and
> staggering in their scope, pace and complexity. Unless we reframe and
> examine them through a social lens, societal responses will be too little, too
> late and potentially blind to negative consequences. (Hackmann et al.
> 2014, p. 653)

The latter half of the twentieth century was witness to enormous social,
cultural and technological developments associated with accelerated globalization, and reflected in, for example, population growth, the expansion of agribusiness, a deepening internationalization of production and
consumption, cultures of consumerism, the development of nuclear
capabilities, the advance of computer and digital technology, and the
ascendancy of the automobile (e.g. Hannigan 2014; Ritzer and Dean
2014). The same period was liberally peppered with significant events that
brought into focus and unsettled the interrelationship between human
and nonhuman nature at the heart of such rapid change: accidental oil
spills such as Torrey Canyon (1967), Ixtoc I (1979), Exxon Valdez (1989)

© The Author(s) 2016
M. Adams, *Ecological Crisis, Sustainability and the Psychosocial Subject*,
DOI 10.1057/978-1-137-35160-9_3

and many others; the Kuwait oil fires and lakes (1991); Windscale, Three Mile Island and Chernobyl nuclear accidents; the Apollo 17 photographs of Earth from space (1972); the sinking of Greenpeace ship *Rainbow Warrior* (1985); and the discovery of the Antarctic 'ozone hole' (1985).[1]

The same period saw the advance of scientific measurement, understanding and analysis of anthropogenic effects described in the previous chapter, encapsulated in the activities and publications of international organizations (e.g. Meadows et al. 1972; Brundtland Report 1987); of analysis and commentary in the social sciences and the humanities (e.g. Commoner 1971; Gorz 1979; Marcuse 1964; Roszak 1992) and of environmentalism as a formal movement, including the creation of many organizations at the forefront of environmental campaigning today (e.g. World Wildlife Fund, formed in 1961; Friends of the Earth 1969; Greenpeace 1971). In the same few decades, numerous government agencies were formed (e.g. US Environmental Protection Agency in 1970; UK Department of the Environment in 1970; Canada Department of the Environment in 1985; China National Environmental Protection Agency in 1984); and government acts were enshrined aimed at addressing environmentalist concerns such as clean air, endangered species and water pollution.

There was also a clear internationalisation of environmental concerns and a fledgling 'sustainable development' agenda during this period, most evident in organizations derived from the United Nations. These have included the World Meteorological Organization (1951), the United Nations Environment Programme (1972), and subsequently the Intergovernmental Panel on Climate Change (1988). Since the first major United Nations Conference on the Environment (1972), the UN has played a central role in asserting the need for action, without having the power or authority for compelling nation states to do so. The UN Conference on Environment and Development (UNCED) or 'Earth Summit' in Rio de Janeiro in 1992 formalized a sustainable development agenda. The outcome of the summit was the ratification of the United Nations Framework Convention on Climate Change (UNFCCC), although all commitments made were non-binding. Signatories agreed to legally binding emissions in the Kyoto Protocol (1997), although a number of countries, including the USA, did not adopt the convention.

In line with the IPCC report findings, the 'ultimate objective' of the treaties adopted 'is to stabilize greenhouse gas concentrations in the atmosphere at a level that will prevent dangerous human interference with the climate system' (UNFCCC 2015). The controversial 2009 United Nations Climate Change Conference in Copenhagen created a further non-binding document—the Copenhagen Accord (UNFCCC 2009). The Accord reiterated the urgency of acting on climate change, and stated that the aim of these actions should be to keep any temperature increases to below 2°C. The summit was subjected to wide-ranging criticism: in terms of a lack of political will and the emphasis on 'voluntary pledges' rather than binding agreements; whether this was likely to be a safe level; and whether it was still a realistic possibility (Müller 2010). At the time of writing, the United Nations Climate Change Conference in Paris in 2015 was considered a partial success for securing broad agreement that the limit of any acceptable rise should be kept to 1.5°C.

Communicating the Anthropocene

While the ostensible purpose of the IPCC and similar bodies has been to gage scientific consensus on the scope and extent of anthropogenic global change, the importance of *communicating* that knowledge, and doing so *effectively* has become a growing concern. This is not surprising considering the recent events and social history as described above. While heads of state and senior policymakers might have been the original target audience of bodies like the UN, getting the message across to a global citizenry is seen as progressively important. This is because a collect global response, however divergently understood, is increasingly acknowledged as necessary to have a hope of mitigating and adapting to its consequences.

As noted above, knowledge and communication of interrelated ecological crises, and their urgency, has led to the rise of various non-governmental organizations, councils, political parties, government bodies and celebrities attempt to communicate the extent of ecological crises more widely.[2] Central amidst this activity is the use of various techniques to attempt to persuade a significant proportion of the populations of wealthy nations to change their behaviour. The nature of these modifications reflects what are perceived to be more or less sustainable behaviours, including energy conservation, transportation, waste and consumption habits (Mazur 2011; Monroe 2003; Moser and Dilling 2007). It is noticeable how discussion of a suitable and significant response to knowing about ecological crisis is often seamlessly folded into discussions of 'sustainable behaviour change', yet what behaviour is relevant here, and what we are trying to sustain, is often ambiguous and/or highly contested (Khoo 2013). The idea of more or less 'sustainable behaviour' is germane to a discussion of how ecological crisis is communicated, and how individuals, households and communities are expected to respond, so it is worth exploring a little more critically.

What emerges then, whether from within the scientific community, or without from environmentalist organizations and commentators, is communication inseparable from attempts to persuade, convince and cajole— to encourage awareness, but also action. With this elision in mind, there is a common way of recounting the history of the communication of the reality of ecological crisis as it has developed post-WWII. It is to describe a journey that begins with naivety, on the part of science communicators and environmentalists, about the way human beings are likely to react to this knowledge (Corner 2014). Naivety stems from the preconception that 'a lack of information and understanding explains the lack of public engagement, and that therefore more information and explanation is needed to move people to action' (Moser and Dilling 2011, p. 162). This is a specific manifestation of the broader adoption of an 'information-deficit model' in the public communication of science (Bak 2001; Sturgis and Allum 2004), whereby 'scientists tend to adopt the information-deficit approach, and to deliver more facts (or the same facts more slowly), assuming that information drives understanding and acceptance' (Rapley et al. 2014, p. 101). It is also apparent in much reporting and campaigning, policy and education programmes (Bulkeley 2000; Shove 2010b). In many, the model is employed intuitively or ad-hoc, rather than an as an explicit theoretical commitment (Nisbet and Scheufele 2009).

It is considered naive because even a modest understanding of developments in psychology and the social sciences suggests that human beings respond in complex ways to knowledge, especially when they have a personal stake in the issue; and that acceptance is, in itself, no guarantee of action (Stokols 1995; Moser and Dilling 2011). As a result, the relevance of the information-deficit model to public science communication has been questioned by social scientists for some time (e.g. Wynne 1992), and in relation to climate change communication in particular (Bulkeley 2000; Uzzell 2000). The problem is often represented as a 'gap'—between attitude and behaviour or values and action (Blake 1999; Boulstridge and Carrigan 2000; Juvan and Dolnicar 2014; Kollmuss and Agyeman 2002; Shove 2010b). There are parallel interdisciplinary calls for more complex models that include a range of psychological, social and cultural contexts that shape understanding and behaviour (e.g. Gifford 2011; Lorenzoni et al. 2007; Kollmuss and Agyeman 2002; Steg and Vlek 2009).

Ecological crisis has clearly not gone away in the last half-century. The science has been further established, environmentalist campaigning and action has become more sophisticated, and the 'sustainability' agenda has gained mainstream political and corporate currency. However, many scholars of the human response to date describe it as largely inadequate in terms of averting crisis for the benefit of current and future flourishing of human life (e.g. Rapley 2012). Initial faith invested in the communication of the science of climate change, or in the power of admonishment and fear to warrant an appropriate collective response (whatever the ambiguities inherent in what we might consider 'appropriate') has dissolved. Faith unpaid, more complex psychological models have been subsequently drawn upon to understand and intervene in what are hoped to be more productive ways. Yet, as I will now argue, the application of psychology, inseparable from shifts in society, culture and political governance, does not provide the solutions we might be encouraged to expect.

Psychology and Behaviour Change

The contemporary application of psychology to the relationship between human beings and the natural environment has many antecedents (e.g. Searles 1960), as do more specific concerns in terms of 'environmentally significant behaviour' (ESB) (Arbuthnot 1977; Liere and Dunlap

1978; Geller et al. 1982; see also Hines et al. 1987 for a meta-analysis). In the early 1990s psychology began to gain prominence as a tool for understanding how and why people behave more or less sustainably (e.g. Deikmann and Preisendorfer 1998; Gardner and Stern 1996; Newhouse 1990; Stokols 1995). If earlier work was concerned with 'the systematic use of applied behavioural analysis' (Willems and McIntire 1982, p. 191), the 1990s saw a broader, more eclectic application of psychology (e.g. Bonaiuto et al. 1996; Fransson and Gärling 1999; Gardner and Stern 1996), including social psychology (e.g. Bonaiuto et al. 1996; McKenzie-Mohr 2000).

Reflecting the focus of international attention encapsulated by the Rio summit in 1992, the 1990s also witnessed a steady increase in the amount of published work dealing with environmental issues across the social sciences, including psychology —although it lagged behind disciplines such as economics and geography (ISSC/UNESCO 2013b, p. 11). But it is during the early twenty-first century that we see the most rapid expansion in the application of psychology to environmental and sustainability issues, and with it increased legitimacy and authority for the models of human behaviour it advances. This growth is evident not only in increased publications and academic profile (e.g. Swim et al. 2011a), but also in the emphasis on psychology in relevant research funding schemes (e.g. Economic and Social Research Council 2015b); in the explicit focus on 'environmental problems' and behaviour change now found in psychological associations in the US and Europe (American Psychological Association 2009; British Psychological Society 2015); and in the take-up of psychological knowledge and research methods by governments and other large organizations (e.g. Crompton 2008; DEFRA 2008). At the time of writing, towards the end of the second decade of the twenty-first century, psychology is commonly utilized to answer questions about how people's attitudes, values, motives and beliefs operate as variables that shape environmentally significant behaviour, and to inform experiments, interventions and policies aimed at encouraging sustainable behaviour and/or reducing unsustainable behaviour (e.g. House of Lords 2011; PlanLoCal 2013; UK Government 2011).[3]

In the rise of psychological explanations, the social and political problems inherent in a focus on individual behaviour change have come into critical focus (Webb 2012; Uzzell and Räthzel 2009). The IPCC have undoubtedly engaged with explanations stressing the behavioural, psychological *and* sociological processes involved in climate change mitigation with increasing depth. However, when it comes to the value of psychology, there is still a tendency, at least in headline statements and Summaries for Policy Makers, to repeat assertions that highlight individual behaviour and assume that changing it just requires more and better application of behavioural psychology and persuasion:

> Emissions can be substantially lowered through changes in consumption patterns (e. g., mobility demand and mode, energy use in households, choice of longer-lasting products) and dietary change and reduction in food wastes. A number of options including monetary and non-monetary incentives as well as information measures may facilitate behavioural changes. (Edenhofer et al. 2014, p. 19)

There is a real issue here in that the Summaries for Policy Makers of each of the Working Groups is what most ministers, policymakers, journalists and interested publics will read. Summary for Policy Maker statements often jar with the much more thorough engagement with social science, even critical social science perspectives, elsewhere in IPCC documentation, such as Working Group reports.[4] The 'options' presented here offer little advance on the behavioural psychology promoted in the 1970s, and ignore the critical reception that has met a return to 'behavioural economics' in recent times (e.g. Jones et al. 2013; Leggett 2014; Mols et al. 2015; Selinger and Whyte 2011). The consequence is that the Summaries effectively obscure the extent to which the detail of IPCC well-informed documentation speaks to power.

Six Problems for a Psychology of Environmentally Significant Behaviour

The limits of addressing human responsibility for ecological crisis in terms of psychological barriers and behavioural interventions aimed at targeting them can be summarized in terms of the following six problems.

(1) Underestimating the Nature and Scope of Change Required

The first, underestimating the nature and scope of change required, is reflected in the tendency to study and promote incremental behavioural change, i.e., 'small steps'; and to focus on working *within* current consumption practices (Swim et al. 2009, p. 144; Willems and McIntire 1982). Critics such as Thøgersen and Crompton (2009) argue that such 'simple and painless' lifestyle change measures are highly unlikely, even if they could be combined, to create the kinds of reduction in demand for energy, consumer goods, forms of transport etc. envisaged by the IPCC to reduce greenhouse gas emissions to the level necessary to keep warming below the 1.5°C threshold (see also Crompton 2013; MacKay 2008; Seyfang 2005; Warde 2013). As Webb puts it, 'in practice, the outcomes of this elaborate behavioural technology, with its complexity, detailed data and refinements, are very conservative' (Webb 2012, p. 112). These are the kinds of manageable, measurable interventions driven by policy agendas, but the policy they inform 'is necessarily incapable of conceptualising transformation in the fabric of daily life on the scale and at the rate required' (Shove 2010b, p. 1283).

Uncritical advocacy of psychology in this area contributes to an understimation of the scope of change required, evident here in Swim et al.'s call (2009, p. 15) for the use of psychological methods: 'Through behavioral-investigations employing experimental and non-experimental methodologies, psychologists can identify the actual determinants of energy consumptive behaviors, many of which are psychological in origin, and can highlight them in communication campaigns to encourage people to behave in more sustainable ways and to promote energy conservation'. The idea that psychological methods can be used to identify 'actual determinants of energy consumptive behaviours', and that they 'are psychological in origin', suggests that such behaviours exist as discreet variables. The notion that communication campaigns can then 'highlight them', which will in turn encourage behaviour change reiterates their psychological separateness, and affirms the sense that they are easily manipulated by 'communication campaigns'. Psychological reductionism accordingly underestimates the systemic level of change demanded by a sociological agenda (Urry 2010); and we might add, fails to attend to the relationships *between* different components and contexts (Capstick et al. 2014). Such attention is precisely what is demanded by a *psychosocial* agenda.

② Depoliticizing the Ecological Crisis

Second, in focusing on individual behaviour change, psychological models of environmentally significant behaviour depoliticizes human responsibility for the ecological crisis (Hobson 2004; Maniates 2001). By emphasizing how unsustainable behaviours can be modified, erased, and how new ones can be encouraged and maintained, broader questions that are more political in nature are effectively sidelined. Such an emphasis also perpetuates a positivist epistemology—'the belief that socially and politically neutral, objective, facts about behaviour can be revealed by scientific technique, and used instrumentally to achieve policy goals' (Webb 2012, p. 112). Associated models of behaviour therefore place emphasis on 'working on and persuading individuals to adapt to a given form of society which in itself is never defined, let alone questioned in any depth' (Uzzell and Räthzel 2009, p. 344). Unasked questions include ones about the sustainability of the production infrastructure in which consumption is located; the social and cultural fabric in which consumerism is swathed; the role and responsibility of governments; the possibilities and potential for political participation and voice; or the activities of powerful incumbent actors with a vested interest in maintaining the status quo (Geels 2014; Paterson and Stripple 2010).

Apolitical accounts of sustainability are exacerbated by the particular ways 'the social' is depicted in prominent discourses (Castree 2014; Castree et al. 2014; see also Leyshon 2014). The 'social' phenomena considered significant tend to be constructed as empirically observable 'objects' such as values and norms, described in terms of factors and variables, and the role of the social sciences more generally as *complimentary* rather than a challenge to scientific worldview – an 'anemic conception' of what the social sciences (and humanities) have to offer according to Castree, that reflects 'a clear unwillingness to unsettle the economic and political status quo' (2014, p. 1). The problem here is what is missing in accounting for 'the social'. For Castree, 'what is screened out in all of this is much of what drives critical social science—such as a focus on power inequalities, violence, and struggle among different constituencies—and much of what preoccupies the humanities—such as the ideas of duty, care, respect, responsibility, rights, faith, cruelty, beauty, and so on' (Castree 2014, p. 1)

③ Ignoring the Power of Conflicting Interests

Third, any attempt to produce incentives, manipulate norms, choices or any other variables imagined to be psychologically or socially distinct 'factors' in real-life settings is likely to find itself in tension with other powerful, often contrary, variations of these attempts. Let us assume for a moment, ethics aside (!) that we take the behaviourist game seriously, and treat culture as a laboratory in which we are attempting to condition human responses towards a sustainable solution. Now imagine the range of punishments, positive and negative reinforcers we would be in competition with, offering alternative messages, promises, incentives, prohibitions, rewards and disincentives. If you find behaviourist terminology objectionable, consider instead the number of norms, narratives, discourses, and in-group and out-group identifications. Corporations, organizations, advertisers, as well as immediate others, are all in the business of promoting certain response behaviours, discouraging others. In the claims made on our behaviour, we find ourselves confronted by a 'surplus of culture' rather than a deficit of information (Hamilton 2012); 'too many confusing signals and meta-signals to react' (Cohen 2013, p. 211). Nonetheless, a particular tension does emerge in attempting to promote sustainable behaviour narratives, especially when they originate in government. This is the tension between sustainability and the continued emphasis on economic growth and consumerism as legitimate and desirable measure of progress (Webb 2012, p. 115). One outcome of this dissonance is to reduce the seriousness of the problem, which relatedly provides us, as consumers, producers, government officials, with a means to disavow the gravity of ecological crisis. As a consequence, we are implicitly permitted to ignore sustainability campaigning, however, limited in the first place.

Reifying Citizens as Passive Subjects

A fourth problem is that psychological models of environmentally significant behaviour encourage narrow constructs of citizenship that rarely exceed the exercise of consumption choices. Whether we understand the reasoning behind sustainability drives is largely inconsequential, unless a 'need-for-cognition' or similar is itself considered an important variable in driving sustainable behaviour. What we think and feel and share about ecological crisis does not seem to matter at all. This conceptualization of agency and autonomy, or rather a lack of it, does not do justice to the extent to which the ecological crisis presents us with a profoundly unsettling set of collective problems, which we are affected by, in ways we are both conscious and unaware of (Soron 2010).[5] Any genuine collective movement must be built on 'the role of humans as reflexive and creative agents of deliberate change'. (ISSC/UNESCO 2013a, b, p. 8).

Fixing Behaviour in Stasis

Fifth, a psychological and behavioural focus too readily fixes in stasis the processes involved in facilitating or preventing environmentally significant behaviour, whether they are considered psychological and/or social. We are routinely presented with a range of variables that psychologists or policymakers are encouraged to utilize in the push for sustainability. Take this example from Swim et al. (2009, p. 138):

> Psychology can help develop both descriptive models and models of behavioral change. Descriptive models of individual behavior delineate the role of internal factors (e.g., knowledge, feelings, values, attitudes) and external factors (e.g., physical and technological infrastructure; political, social and cultural factors; economic incentives; social influences and models) in environmentally significant behaviour... Descriptive models of behavior in groups may explicate the conditions under which groups will or will not provide public goods or fall prey to the commons dilemma.

This is an ostensibly comprehensive list, but there is no suggestion that any of these factors are anything other than fixed entities, and the implication is that they interact with each other in ways that are potentially predictable, and therefore controllable. This is appealing and hopeful, for it creates a sense of control. However, the fluidity of these 'factors' in the context of everyday phenomenal experience, their interpenetration and mutual contingency punctures this sense of mastery. For a psychosocial perspective, attitudes, behaviours and feelings described as 'internal factors' are always intertwined with what are listed as 'external factors' – 'embedded in the flux of social relations and structures from which they derive meaning' (Webb 2012, p. 117).

In practice, this level of interrelatedness suggests that the role of psychology is too readily overstated, because human reactions are more complex and unpredictable than what factors and variables models suggest. But it might also explain why concerned natural scientists uncritically encourage policymakers to utilize 'behavioural science', e.g.: 'Government policies are needed when people's behaviors fail to deliver the public good. Those policies will be most effective if they can stimulate long-term changes in beliefs and norms, creating and reinforcing the behaviors needed to solidify and extend the public good' (Kinzig et al. 2013, p. 164).

The psychologization of sustainability also contributes to an immobilization and sequestration of the nonhuman natural environment. It is ironic that although the object of all of these models is behaviour that has a significant impact on 'the environment', what role that 'environment' has in shaping experience and behaviour is completely absent; including how it contributes to other social and psychological factors, and is permeated and constituted by them. The 'environment' is to be saved, conserved, but it still appears as an external object, a painted backdrop upon which more complex human affairs are manifest. The idea of a more fluid human-nonhuman encounter, of interweaving ecological, cultural, biological and social processes, that we see emerging in anthropology and other disciplines (Kohn 2013; Latour 2005), still appears beyond the range of convenience for mainstream psychology.

Neglecting the Importance of Social Context

A sixth and final issue is in part a culmination of previous points— a neglect of the importance of social context. A focus on psychological barriers to pro-environmental behaviour distracts from the social, structural and cultural contexts in which environmentally significant behaviours are embedded. The salient feature of a range of critiques here is the claim 'that social context has been under theorized and researched in psychological work on environmental problems and solutions, resulting in individualistic understanding of human practices and misguided policy interventions assuming people behave as rational agents' (Adams 2012, p. 217; Szerszynski and Urry 2010). It follows then that these contexts *also* shape that behaviour, and must be considered as barriers or enablers of environmentally significant behaviour alongside psychological ones. The next section considers to what extent the social sciences are addressing such contexts in the study of ecological crisis and sustainability.

Out of the Shadows?

Despite the criticisms just discussed, there is a noticeable 'social turn' in the way a range of disciplines, including some areas of psychology, approach the causes and consequences of ecological crisis, and formulations of how we might successfully 'mitigate' or 'adapt' to its effects via sustainable development. We see some signs of this in the more inclusive psychological models discussed above. Lowe et al. go as far as to say that it has become a 'mantra' in contemporary science policy on environmental issues to at least call for engagement with 'a wide range of sciences' including the social sciences (2013, p. 207). After what is commonly perceived to be a slow start (Kasper 2009), mainstream sociology and the social sciences are also paying increasing attention to ecological crisis, sustainability issues and the human-nonhuman relationship (e.g. ISCC/UNESCO 2013b). As a result, the incorporation of social science knowledge into scientific programmes and non-governmental organization activity is on the rise, however partial and ambivalent this process (Castree 2014). While some accounts of 'the social' advanced or utilized here carve it up into discreet variables, subject to many of the limitations described above, there exists 'shadow, subordinated constructions of public issues, which convey a more complex analysis of the social relations of consumption and its political-economic determinants, but this seems to be outside what can be openly acknowledged and debated in formal policy commitments' (Webb 2012, pp. 120–121).

Since Webb made this claim, there is evidence that a more explicit engagement with social perspectives is emerging out of the shadows. Hackmann et al. go as far as to say that 'the debate on global environmental change is shifting from a predominant focus on biophysical processes to a focus on societal processes and concerns interacting with the climate and environment' (Hackmann et al. 2014, p. 653). This is a bold statement, with undoubted appeal for social scientists who have for some time stressed the need to address social context in understanding ecological crisis and sustainable development (e.g. Kidner 2012; 2001; Uzzell 2000). It begs at least two further questions: Where and how is the shift in focus discernable? And how are 'societal processes and concerns' defined and delineated?

In answering the first question, greater emphasis on socio-cultural dynamics is clearly finding its way into some level of scientific and policy analysis (European Commission 2010; Scottish Government 2010); including the most recent IPCC assessment process and subsequent report (AR5), especially the contributions of Working Groups II and III (Field et al. 2014; Edonhofer et al. 2014). In these reports there is a discernible, often explicit, reframing of global environmental change as a social process (Hackmann et al. 2014). Another significant collective publication (150+ authors) is the ISSC/UNESCO *World Social Science Report* (WSSR) on global environmental change (ISSC/UNESCO 2013a, b). According to a number of its contributing authors, the *WSSR* is an explicit articulation of the implied social reframing we find in the IPCC Working Group reports (Weaver et al. 2014). In these three reports alone we find a remarkable amount of recent scholarly work collating, distilling and integrating social science perspectives on all aspects of ecological crisis and sustainable development. Adding to this an increasing call amongst geoscientific communities for greater links with the social sciences and some attempts to make those links (though see Castree et al. 2014 for critique); and ambitious new international and interdisciplinary research 'hubs' such as Future Earth that build on earlier programmes.[6]

While organizations like the IPCC are belatedly placing 'the social at the heart of global environmental change' (Hackmann et al. 2014), they are inevitably playing catch up with how existing humanities and social science theory and research have addressed these issues, and more specifically, the contribution of an avowedly psychosocial perspective. Relatedly, a clarion call for the social sciences to help 'fundamentally reframe climate and environmental change from a physical into a social problem' (Hackmann et al. 2014), means making use of existing reframing work, that for whatever reason, has been ignored or marginalized. One of the tasks of this book to explore the gap between representations of a social science agenda in reports such as AR5 and the breadth and depth of social science accounts more widely. A related task is to examine how 'the social' is articulated in the shift in the official narrative represented by AR5 and other sources; and how it compares to existing and emerging social science perspectives. Of course this is in some ways an impossible task – there is no privileged position from which to form an objective overview of these areas, or the gap between them. But as someone who has been immersed in social science literature addressing global environmental change for the last few years, and now researching mainstream accounts of social science narratives, I can at least offer an account.

Answering the second question – how are societal processes and concerns defined and delineated—is complicated by the breadth and variety of theory and research covered, even if we restrict ourselves to the documentation noted above.[7] The basic aim of the *World Social Science Report* is to envision a more central role for social scientists, as part of a 'global science community'. Accordingly they encourage social scientists to be pro-active promoters of the value of social science perspectives for understanding global environmental change; and exhort policymakers, funding bodies, research teams and other key organizations to recruit and prioritize social science expertise in projects and priorities from the outset.[8] They reflect an optimism that social scientific knowledge has great potential (Shove 2010c), and will be welcome: 'The field is growing, wide open, and rife with opportunity to broaden and deepen what social scientists can contribute to the topic of global environmental change and sustainability' (UNESCO/ICSS 2013, p. 6).

What is it that social scientists are expected to bring to the table as 'an essential piece of the research puzzle' (Hackmann and St. Clair 2012, p. 7)? The IPCC Working Group III report provides a whole chapter addressing 'the strengths and limitations of the most widely used concepts and methods in economics, ethics, and other social sciences that are relevant to climate change' (Kolstad et al. 2014, p. 211); and another on sustainable development (Fleurbaey et al. 2014). Kolstad and colleagues identify the main contribution of the social sciences as deriving from the UNFCCC key objective: 'to avoid dangerous anthropogenic interference with the climate system'. There are inherent ambiguities in this objective that require social judgement:

> Two main issues confronting society (and the IPCC) are: what constitutes 'dangerous interference' with the climate system and how to deal with that interference. Determining what is dangerous is not a matter for natural science alone; it also involves value judgements —a subject matter of the theory of value, which is treated in several disciplines, including ethics, economics, and other social sciences. (Kolstad et al. 2014, p. 211)

'Value judgements' are concerned not just with the level of mitigation needed but also how to go about it, which involves big ethical and political questions: who is most responsible, and how mitigation efforts should be distributed globally and temporally incorporates a wide range of issues about what amounts to dangerous, and what should be done about it. While international treaties have largely failed to date, the hope here is that by explicitly acknowledging the human processes involved, social science knowledge can help nurture social processes and norms, customs, communities, institutions and cultures, required for sustainable development; promoting trust, co-operation, altruism, reciprocity, distributive justice and suitable structures for acting collectively (Kolstad et al. 2014, pp. 213–254). Oddly, despite the broad and ambitious role allocated to social understandings here, in terms of gaps in knowledge and data that need to be addressed in relation to mitigation, the social sciences are not mentioned at this juncture. Instead, only 'more research that incorporates behavioural economics into climate change mitigation is needed' (Kolstad et al. 2014, p. 258).

Analysis of the role of 'the social' in sustainable development comes later in the report (Fleurbaey et al. 2014). Here, there is a clearer acknowledgment of the need to address social context. We find the usual habit of listing a series of factors as if they were easy to identify and separate out, but there is also an attempt to acknowledge the complex situatedness of these factors. In a discussion of the possibility of reducing or shifting consumption patterns in areas such as household energy demand, for example, this description is offered:

> 'consumption practices and patterns are influenced by a range of economic, informational, psychological, sociological, and cultural factors, operating at different levels or spheres in society – including the individual, the family, the locality, the market, and the work place... Furthermore, consumers' preferences are often constructed in the situation (rather than pre-existing) and their decisions are highly contextual... and often inconsistent with values, attitudes, and perceptions... sustainable consumption practices are bound up with perceptions of identity, ideas of good life, and so on. (pp. 307–308)

Consumption behaviour, significantly described as 'practices and patterns' rather than individual acts, are presented here as sites of a double intersection. They are influenced by a range of factors—economic, informational etc.—but these, in turn, combine in different ways as they operate in different spheres of everyday life—situated, contingent and contextual processes. Fleurbay et al.'s description marks a more integrated, fluid and dynamic account, ingredients of what Capstick et al. outline as a 'radical social science' agenda (Capstick et al. 2014). Similarly, the *World Social Science Report* identifies the 'emerging story as:

> ... one of individuals richly and dynamically embedded in households, communities, sociotechnical systems, economies and cultures. It goes a long way toward explaining the paradox of how the social drivers of global environmental change persist, or at least change only slowly, while environmental crises continue to unfold rapidly. (ICSS/UNESCO 2013, p. 17)

There appears to be hope that an emerging critical social science agenda will reject a focus on 'pro-environmental behaviour', small steps campaigns, the psychologization of the sustainability, behavioural economics and the behaviour change agenda (Thøgersen and Crompton 2009). More constructively, an ambitious and radical social science perspective points to the need to integrate multiple approaches, and calls for an 'in-depth consideration and understanding of the complexity and influence of the social, economic contexts in which [observable behaviours] manifest, arise and develop' (Capstick et al. 2014, p. 8). A psychosocial agenda would also highlight the ways more complex dynamics of affect, embodiment and psyche are interwoven with those contexts; and I would further add that human relationships with, and embeddedness in, nonhuman nature must be considered integral to this mix.

The rest of this book is an attempt to further the conceptualization of these interrelationships, taking in theories and research from across a range of disciplines including anthropology, cultural studies, ethology, psychoanalysis, sociology and social psychology. We begin with social practice theory, which is influenced by the work of Bourdieu and Giddens amongst others. The theory takes practice, rather than observable behaviour, as its basic unit of analysis (Shove 2010b). A practice is understood as a 'way of doing', rather than a discreet behaviour. The way things are done depends upon the ongoing amalgam of interconnecting material, cultural and individual dynamics (Shove et al. 2012). It follows that social practice theory has the potential to offer a more coherent, integrated and comprehensive account of the 'drivers' of sustainable behaviour (Scheele and Papazu 2015). Compared to prevailing behaviour change orthodoxies, social practice theory also points to a more ambitious articulation of the scope of change required to effect meaningful emission reductions (Capstick et al. 2015). Accordingly, the theory has been taken up enthusiastically in recent years to conceptualize sustainable behaviour, and to a lesser extent, inform interventions (e.g. Hand et al. 2005; Hargreaves 2011b; Spotswood et al. 2015). In light of this growing interest, a number of authors have asked whether contemporary versions of social practice theory do indeed meet the requirement for an integrated approach to sustainability and social change, successfully addressing the multiple contexts involved (Adams 2012; Capstick et al. 2014; Kurz et al. 2015). The next two chapters consider whether the promise of social practice theory is borne out by closer scrutiny.

Notes

1. The twenty-first century has continued in similar fashion, already registering the largest accidental oil spill of all time (Deepwater Horizon 2010); the Fukushima Daiichi nuclear disaster (2011), and extreme weather events including the Boxing Day Tsunamis (2004), Hurricane Katrina and others (2005).
2. Anderson (2011) and Boykoff and Goodman (2009) offer perceptive and entertaining analyses of the pitfalls of celebrity endorsement of environmental issues.
3. An illustrative example of psychology's moment in the sustainability spotlight is the activity of the largest and most powerful psychological association in the world – the American Psychological Association. The Association created a climate change 'task force', which produced an extensive report (American Psychological Association 2009), followed by later updates (Doherty and Clayton 2011; Swim et al. 2011a, b; Reser and Swim 2011; Gifford 2011; Stern 2011).
4. After reading these documents in some detail, it is easy to concur with Black's more general criticism that 'the IPCC has shown a remarkably consistent capacity to turn out documents that defy comprehension by the non-specialist, despite the undoubted quality of the underlying assessments' (Black 2015, p. 282).
5. This shared context is presumably what motivates natural and social scientists concerned about anthropogenic ecological degradation to commit themselves to developing relevant theoretical models, carry out research and communicate their findings.
6. See www.futureearth.org. See also the International Council for Science's International Human Dimensions Programme (IHDP est. 1996); and the International Geosphere-Biosphere Programme (IGBP, est. 1987).
7. The IPCC Working Group Reports alone amount to over 4000 pages.
8. Of course persistent calls for a heightened role for social science perspectives is not itself evidence of a greater presence for social science understandings, let alone critical ones. The actual influence of social science perspectives on policy formation, international treaties, and collective responses more generally is yet to be seen.

References

Adams, M. (2012). A social engagement: How ecopsychology can benefit from dialogue with the social sciences. *Ecopsychology, 4*(3), 216–222.

American Psychological Association Task Force on the Interface Between Psychology and Global Climate Change. (2009). *Psychology and global climate change: Addressing a multi-faceted phenomenon and set of challenges.* Available at http://www.apa.org/science/climate-change/

Anderson, A. (2011). Sources, media, and modes of climate change communication: The role of celebrities. *WIREs Climate Change, 2*, 535–546.

Arbuthnot, J. (1977). The roles of attitudinal and personality variables in the prediction of environmental behavior and knowledge. *Environment and Behavior, 9*(2), 217–232.

Bak, H. J. (2001). Education and public attitudes toward science: Implications for the "deficit model" of education and support for science and technology. *Social Science Quarterly, 82*(4), 779–795.

Black, R. (2015, April 5). No more summaries for wonks. *Nature: Climate Change*, pp. 282–284. www.nature.com/natureclimatechange. Accessed 18 Dec 2015.

Blake, J. (1999). Overcoming the 'value-action gap' in environmental policy: Tensions between national policy and local experience. *Local Environment, 4*(3), 257–278.

Bonaiuto, M., Breakwell, G. M., & Cano, I. (1996). Identity processes and environmental threat: The effects of nationalism and local identity upon perception of beach pollution. *Journal of Community and Applied Social Psychology, 6*, 157–175.

Boulstridge, E., & Carrigan, M. (2000). Do consumers really care about corporate responsibility? Highlighting the attitude-behaviour gap. *Journal of Communication Management, 4*(4), 355–368.

Boykoff, M. T., & Goodman, M. (2009). Conspicuous redemption: Promises and perils of celebrity involvement in climate change. *Geoforum, 2009*(40), 395–406.

British Psychological Society. (2015). Behaviour change briefings. http://www.bps.org.uk/what-we-do/our-influence/behaviour-change-briefings/behaviour-change-briefings-0. Accessed 18 Dec 2015.

Brundtland, G. H. (1987). *Our common future: Report of the 1987 world commission on environment and development* (pp. 1–59). Oslo: United Nations.

Bulkeley, H. (2000). Common knowledge? Public understanding of climate change in Newcastle, Australia. *Public Understanding of Science, 9*(3), 313–334.

Capstick, S., Lorenzoni, I., Corner, A., & Whitmarsh, L. (2014). Prospects for radical emissions reduction through behavior and lifestyle change. *Carbon Management, 5*(4), 429–445.

Capstick, S., Whitmarsh, L., Poortinga, W., Pidgeon, N., & Upham, P. (2015). International trends in public perceptions of climate change over the past quarter century. *Wiley Interdisciplinary Reviews: Climate Change, 6*(1), 35–61.

Castree, N. (2014, November 14). Dangerous knowledge and global environmental change: Whose epistemologies count? *EnviroSociety.* www.envirosociety.org/2014/11/dangerous-knowledge-and-global-environmental-change. Accessed 18 Dec 2015.

Castree, N., Adams, W. M., Barry, J., Brockington, D., Buscher, B., Corbera, E., Demeritt, D., Duffy, R., Felt, U., Neves, K., Newell, P., Pellizzoni, L., Rigby, K., Robbins, P., Robin, L., Rose, D. B., Ross, A., Schlosberg, D., Sorlin, S., West, P., Whitehead, M., & Wynne, B. (2014). Changing the intellectual climate. *Nature Climate Change, 4*(9), 763–768. doi:10.1038/NCLIMATE2339.

Cohen, S. (2013). *States of denial: Knowing about atrocities and suffering* (2nd ed.). New York: Wiley.

Commoner, B. (1971). *The closing circle.* New York: Knopf.

Corner, A. (2014). *Climate Silence (and how to break it).* London: Climate Outreach & Information Network. http://climateoutreach.org/resources/climate-silence-and-how-to-break-it/. Accessed 12 Sep 2016.

Crompton, T. (2008). *Weathercocks and signposts.* World Wildlife Fund. www.wwf.org.uk/strategiesforchange. Accessed 21 Dec 2015.

Crompton, T. (2013). Behaviour change: A dangerous distraction. In R. Crocker & S. Lehmann (Eds.), *Motivating change: Sustainable design and behaviour in the built environment* (pp. 111–126). London: Routledge.

Defra. (2008). *A framework for pro-environmental behaviours.* London: Defra.

Diekmann, A., & Preisendorfer, P. (1998). Environmental behavior – Discrepancies between aspirations and reality. *Rationality and Society, 10,* 79–102.

Doherty, T. J., & Clayton, S. (2011). The psychological impacts of global climate change. *American Psychologist, 66*(4), 265–276.

Edenhofer, O., Pichs-Madruga, R., Sokona, Y., Farahani, E., Kadner, S., Seyboth, K., Adler, A., Baum, I., Brunner, S., Eickemeier, P., Kriemann, B.,

Savolainen, J., Schlömer, S., von Stechow, C., Zwickel, T., & Minx, J. C. (Eds.). (2014). *Climate Change 2014: Mitigation of Climate Change. Summary for Policymakers. Contribution of Working Group III to the Intergovernmental Panel on Climate Change, Fifth Assessment Report.* Cambridge: Cambridge University Press.

ESRC. (2015b). *Environment and energy research.* http://www.esrc.ac.uk/research/research-topics/environment-and-energy-research/. Accessed 22 Dec 2015.

European Commission. (2010). *Work Programme 2011, Cooperation Theme 6, Environment (including climate change).* ftp://ftp.cordis.europa.eu/pub/fp7/docs/wp/cooperation/environment/f-wp-201101_en.pdf. Accessed 18 Dec 2015.

Field, C. B., Barros, V. R., Dokken, D. J., Mach, K. J., Mastrandrea, M. D., Bilir, T. E., Chatterjee, M., Ebi, K. L., Estrada, Y. O., Genova, R. C., Girma, B., Kissel, E. S., Levy, A. N., MacCracken, S., Mastrandrea, P. R., & White, L. L. (Eds.). (2014). *Climate change 2014: Impacts, adaptation, and vulnerability. Summary for policymakers. Contribution of Working Group II to the Intergovernmental Panel on Climate Change, fifth assessment report.* Cambridge: Cambridge University Press.

Fleurbaey, M., Kartha, S., Bolwig, S., Chee, Y. L., Chen, Y., Corbera, E., Lecocq, F., Lutz, W., Muylaert, M. S., Norgaard, R. B., Okereke, C., & Sagar, A. D. (2014). Sustainable development and equity. In O. Edenhofer, R. Pichs-Madruga, Y. Sokona, E. Farahani, S. Kadner, K. Seyboth, A. Adler, I. Baum, S. Brunner, P. Eickemeier, B. Kriemann, J. Savolainen, S. Schlömer, C. von Stechow, T. Zwickel, & J. C. Minx (Eds.), *Climate change 2014: Mitigation of climate change. Contribution of Working Group III to the fifth assessment report of the Intergovernmental Panel on Climate Change.* Cambridge: Cambridge University Press.

Fransson, N., & Gärling, T. (1999). Environmental concern: Conceptual definitions, measurement methods, and research findings. *Journal of Environmental Psychology, 19*(4), 369–382.

Gardner, G. T., & Stern, P. C. (1996). *Environmental problems and human behavior.* Boston: Allyn and Bacon.

Geels, F. W. (2014). Regime resistance against low-carbon transitions: Introducing politics and power into the multi-level perspective. *Theory, Culture and Society, 31*(5), 21–40.

Geller, E. S., Winett, R. A., Everett, P. B., & Winkler, R. C. (1982). *Preserving the environment: New strategies for behavior change.* New York: Pergamon Press.

Gifford, R. (2011). The dragons of inaction: Psychological barriers that limit climate change mitigation and adaptation. *American Psychologist, 66*(4), 290–302.

Gorz, A. (1979). *Ecology as politics.* London: Southend Press.

Hackmann, H., & St. Clair, A. L. (2012). *Transformative cornerstones of social science research for global change.* (International Social Science Council). http://www.worldsocialscience.org/documents/transformative-cornerstones.pdf. Accessed 18 Dec 2015.

Hackmann, H., Moser, S. C., & St. Clair, A. L. (2014). The social heart of global environmental change. *Nature Climate Change, 4,* 653–655.

Hamilton, C. (2012a). What history can teach us about climate change denial. In S. Weintrobe (Ed.), *Engaging with climate change: Psychoanalytic & interdisciplinary perspectives* (pp. 16–32). London: Routledge.

Hand, M., Shove, E., & Southerton, D. (2005). Explaining showering: A discussion of the material, conventional, and temporal dimensions of practice. *Sociological Research Online, 10*(2). http://www.socresonline.org.uk/10/2/hand.html. Accessed 18 Dec 2015.

Hannigan, J. (2014). *Environmental sociology* (3rd ed.). London: Routledge.

Hargreaves, D. (2011a). Pro-environmental interaction: Engaging Goffman on pro-environmental behaviour change. *Centre for Social and Economic Research on the Global Environment Working Papers 11-04.* Norwich: University of East Anglia. http://www.cserge.ac.uk/sites/default/files/2011-04.pdf. Accessed 18 Dec 2015.

Hargreaves, T. (2011b). Practice-ing behaviour change: Applying social practice theory to pro-environmental behaviour change. *Journal of Consumer Culture, 11*(1), 79–99.

Hines, J. M., Hungerford, H. R., & Tomera, A. N. (1986/1987). Analysis and synthesis of research on environmental behavior: A meta- analysis. *Journal of Environmental Education, 18,* 1–8.

Hobson, K. (2004). Sustainable consumption in the United Kingdom: The "responsible" consumer and government at "arm's length". *The Journal of Environment & Development, 13*(2), 121–139.

House of Lords. (2011). *Behaviour change.* House of Lords, Science and Technology Select Committee. HL Paper no. 179.

ISSC/UNESCO. (2013a). *SUMMARY: World Social Science Report 2013: Changing global environments.* Paris: OECD Publishing and UNESCO Publishing.

ISSC/UNESCO. (2013b). *World Social Science Report 2013: Changing global environments*. Paris: OECD Publishing and UNESCO Publishing.

Jones, R., Pykett, J., & Whitehead, M. (2013). Psychological governance and behaviour change. *Policy and Politics, 41*(2), 159–182.

Juvan, E., & Dolnicar, S. (2014). The attitude-behaviour gap in sustainable tourism. *Annals of Tourism Research, 48*, 76–95.

Kasper, D. V. S. (2009). Ecological habitus: Toward a better understanding of socioecological relations. *Organization and Environment, 22*(3), 311–326.

Khoo, S. (2013). Sustainable development of what? In F. Fahy & H. Rau (Eds.), *Methods of sustainability research in the social sciences*. London: Sage.

Kidner, D. W. (2001). *Nature and psyche: Radical environmentalism and the politics of subjectivity*. New York: SUNY Press.

Kidner, D. (2012). *Nature and experience in the culture of delusion: How industrial society lost touch with reality*. Basingstoke: Palgrave Macmillan.

Kinzig, A. P., Ehrlich, P. R., Alston, L. J., Arrow, K., Barrett, S., Buchman, T. G., … & Saari, D. (2013). Social norms and global environmental challenges: The complex interaction of behaviors, values, and policy. *BioScience, 63*(3), 164–175.

Kohn, E. (2013). *How forests think: Toward an anthropology beyond the Human*. Berkeley: University of California Press.

Kollmuss, A., & Agyeman, J. (2002). Mind the gap: Why do people act environmentally and what are the barriers to pro-environmental behavior? *Environmental Education Research, 8*(3), 239–260.

Kolstad, C., Urama, K., Broome, J., Bruvoll, A., Cariño Olvera, M., Fullerton, D., Gollier, C., Hanemann, W. M., Hassan, R., Jotzo, F., Khan, M. R., Meyer, L., & Mundaca, L. (2014). Social, economic and ethical concepts and methods. In O. Edenhofer, R. Pichs-Madruga, Y. Sokona, E. Farahani, S. Kadner, K. Seyboth, A. Adler, I. Baum, S. Brunner, P. Eickemeier, B. Kriemann, J. Savolainen, S. Schlömer, C. von Stechow, T. Zwickel, & J. C. Minx (Eds.), *Climate change 2014: Mitigation of climate change. Contribution of Working Group III to the fifth assessment report of the Intergovernmental Panel on Climate Change*. Cambridge, UK/New York: Cambridge University Press.

Kurz, T., Gardner, B., Verplanken, B., & Abraham, C. (2015). Habitual behaviors or patterns of practice? Explaining and changing repetitive climate-relevant actions. *WIREs Clim Change, 6*, 113–128. doi:10.1002/wcc.327.

Latour, B. (2005). *Reassembling the social*. Oxford: Oxford University Press.

Leggett, W. (2014). The politics of behaviour change: Nudge, neoliberalism and the state. *Policy and Politics, 42*(1), 3–19.

Leyshon, C. (2014). Critical issues in social science climate change research. *Contemporary Social Science: Journal of the Academy of Social Sciences, 9*(4), 359–373.

Liere, K. D., & Dunlap, R. E. (1978). Moral norms and environmental behavior: An application of Schwartz's norm-activation model to yard burning. *Journal of Applied Social Psychology, 8*(2), 174–188.

Lorenzoni, I., Nicholson-Cole, S., & Whitmarsh, L. (2007). Barriers perceived to engaging with climate change among the UK public and their policy implications. *Global Environmental Change, 17*(3), 445–459.

Lowe, P., Phillipson, J., & Wilkinson, K. (2013). Why social scientists should engage with natural scientists. *Contemporary Social Science, 8*(3), 207–222.

MacKay, D. (2008). *Sustainable Energy – Without the hot air.* Cambridge: UIT.

Maniates, M. (2001). Individualization: Plant a tree, buy a bike, save the world? *Global Environmental Politics, 1*(3), 31–52.

Marcuse, H. (1964). *One dimensional man. Studies in the ideology of advanced industrial society.* Boston: Beacon Press.

Mazur, L. (2011). Inspiring action: The role of psychology in environmental campaigning and activism. *Ecopsychology, 3*(2), 139–148.

McKenzie-Mohr, D. (2000). Promoting sustainable behavior: An introduction to community-based social marketing. *Journal of Social Issues, 56*(3), 543–554.

Meadows, D. H., Meadows, G., Randers, J., & Behrens, W. W. (1972). *The limits to growth.* New York: Universe Books.

Mols, F., Haslam, S. A., Jetten, J., & Steffens, N. K. (2015). Why a nudge is not enough: A social identity critique of governance by stealth. *European Journal of Political Research, 54*, 81–98.

Monroe, M. C. (2003). Two avenues for encouraging conservation behaviors. *Human Ecology Review, 10*(2), 113–125.

Moser, S. C., & Dilling, L. (Eds.). (2007). *Creating a climate for change: Communicating climate change.* Cambridge: Cambridge University Press.

Moser, S. C., & Dilling, L. (2011). Communicating climate change: Closing the science-action gap. *The Oxford handbook of climate change and society* (pp. 161–174). Oxford: Oxford University Press.

Müller, B. (2010). *Copenhagen 2009: Failure or final wake-up call for our leaders?* Oxford: Oxford Institute for Energy Studies.

Newhouse, N. (1990). Implications of attitude and behavior research for environmental conservation. *The Journal of Environmental Education, 22*(1), 26–32.

Nisbet, M. C., & Scheufele, D. A. (2009). What's next for science communication? Promising directions and lingering distractions. *American Journal of Botany, 96*(10), 1767–1778.

Paterson, M., & Stripple, J. (2010). My space: Governing individuals' carbon emissions. *Environment and Planning. D, Society and Space, 28*(2), 341.

PlanLoCal. (2013). *Behaviour change: Theories, approach, guidance.* Centre for Sustainable Energy. Available via the reading list or at http://www.planlocal. org.uk/pages/getting-people-involved/behaviour-change-theories-approaches-and-guidance

Rapley, C. G. (2012). Foreword. In S. Weintrobe (Ed.), *Engaging with climate change: Psychoanalytic and interdisciplinary perspectives* (pp. xviii–xvxxi). London: Routledge.

Rapley, C. G., De Meyer, K., Carney, J., Clarke, R., Howarth, C., Smith, N., et al. (2014). *Time for change? Climate science reconsidered: Report of the UCL policy commission on communicating climate science.* London: UCL.

Reser, J. P., & Swim, J. K. (2011). Adapting to and coping with the threat and impacts of climate change. *American Psychologist, 66*(4), 277.

Ritzer, G., & Dean, P. (2014). *Globalization: A basic text.* London: Wiley.

Roszak, T. (1992). *The voice of the earth.* New York: Simon and Schuster.

Scheele, C. E., & Papazu, I. (2015). Changing individual behaviors or creating green societies? Advancing from a behaviorist to a social practice theory approach. *Ecopsychology, 7*(2), 104–111.

Searles, H. (1960). *The nonhuman environment in normal development and in schizophrenia.* New York: International Universities Press.

Selinger, E., & Whyte, K. (2011). Is there a right way to nudge? The practice and ethics of choice architecture. *Sociology Compass, 5*(10), 923–935.

Seyfang, G. (2005). Shopping for sustainability: Can sustainable consumption promote ecological citizenship? *Environmental Politics, 14*(2), 290–306.

Shove, E. (2010a). Sociology in a changing climate. *Sociological Research Online, 15*(3), 12 http://www.socresonline.org.uk/15/3/12.html

Shove, E. A. (2010b). Beyond the ABC: Climate change policy and theories of social change. *Environment and Planning. A, 42*(6), 1273–1285.

Shove, E. A. (2010c). Social theory and climate change questions often, sometimes and not yet asked. *Theory, Culture and Society, 27*(2–3), 277–288.

Shove, E. A. (2010d). Submission to the House of Lords Science and Technology. Select Committee Call for Evidence on Behaviour Change. Available at: http://www.lancs.ac.uk/staff/shove/transitionsinpractice/papers/ShoveHouseofLords.pdf

Shove, E. A., Pantzar, M., & Watson, M. (2012). *The dynamics of social practice: Everyday life and how it changes*. London: Sage.

Soron, D. (2010). Sustainability, self-identity and the sociology of consumption. *Sustainable Development, 18*(3), 172–181.

Spotswood, F., Chatterton, T., Tapp, A., & Williams, D. (2015). Analysing cycling as a social practice: An empirical grounding for behaviour change. *Transportation Research Part F: Traffic Psychology and Behaviour, 29*, 22–33.

Steg, L., & Vlek, C. (2009). Encouraging pro-environmental behaviour: An integrative review and research agenda. *Journal of Environmental Psychology, 29*(3), 309–317.

Stern, P. C. (2011). Contributions of psychology to limiting climate change. *American Psychologist, 66*(4), 303–314.

Stokols, D. (1995). The paradox of environmental psychology. *American Psychologist, 50*(10), 821–837.

Sturgis, P., & Allum, N. (2004). Science in society: Re-evaluating the deficit model of public attitudes. *Public Understanding of Science, 13*(1), 55–74.

Swim, J., Clayton, S., Doherty, T., Gifford, R., Howard, G., Reser, J., & Weber, E. (2009). *Psychology and global climate change: Addressing a multi-faceted phenomenon and set of challenges. A report by the American Psychological Association's task force on the interface between psychology and global climate change*. Washington, DC: American Psychological Association.

Swim, J. K., Stern, P. C., Doherty, T. J., Clayton, S., Reser, J. P., Weber, E. U., Gifford, R., & Howard, G. S. (2011a). Psychology's contributions to understanding and addressing global climate change. *American Psychologist, 66*(4), 241.

Swim, J. K., Clayton, S., & Howard, G. S. (2011b). Human behavioral contributions to climate change: Psychological and contextual drivers. *American Psychologist, 66*(4), 251–264.

The Scottish Government. (2010). *10 Key messages about behaviour change*. http://www.scotland.gov.uk/Topics/Research/by-topic/environment/social-research/Remit/events/Key-Messages. Accessed 18 Dec 2015.

Thøgersen, J., & Crompton, T. (2009). Simple and painless? The limitations of spillover in environmental campaigning. *Journal of Consumer Policy, 32*, 141–163.

UK Government. (2011). *Cabinet Office Behavioural Insights Team: Behaviour Change and Energy Use*. http://www.cabinetoffice.gov.uk/resource-library/behaviour-change-and-energy-use. Accessed 18 Dec 2015.

UNFCCC. (2009, December 18). Copenhagen Accord. *U.N. Framework Convention on Climate Change*. United Nations. http://unfccc.int/meetings/copenhagen_dec_2009/items/5262.php. Accessed 21 Dec 2015.

UNFCCC. (2015). About UNFCCC. http://newsroom.unfccc.int/about/. Accessed 18 Dec 2015.

Urry, J. (2010). Sociology facing climate change. *Sociological Research Online, 15*(3), 1.

Uzzell, D. L. (2000). The psycho-spatial dimension of global environmental problems. *Elsevier Journal of Environmental Psychology, 20*(4), 307–318.

Uzzell, D., & Räthzel, N. (2009). Transforming environmental psychology. *Journal of Environmental Psychology, 29*(3), 340–350.

Warde, A. (2013) Sustainable consumption and behaviour change. *Discover Society*, Issue 1. http://www.discoversociety.org/2013/10/01/sustainable-consumption-and-behaviour-change/. Accessed 12 Sep 2016.

Weaver, C. P., Mooney, S., Allen, D., Beller-Simms, N., Fish, T., Grambsch, A. E., & Winthrop, R. (2014). From global change science to action with social sciences. *Nature Climate Change, 4*(8), 656–659.

Webb, J. (2012). Climate change and society: The chimera of behaviour change technologies. *Sociology, 46*(1), 109–125.

Willems, E. P., & McIntire, J. D. (1982). A review of preserving the environment: New strategies for behavior change. *The Behavior Analyst, 5*(2), 191–197.

Wynne, B. (1992). Misunderstood misunderstanding: Social identities and public uptake of science. *Public Understanding of Science, 1*(3), 281–304.

4

Searching for a New Normal: Social Practices and Sustainability

Introduction

> What becomes possible in terms of imagining behaviour change… once the assumption of a sovereign subject is suspended, embodiment foregrounded, and habit re-imagined? (Schwanen et al. 2012, p. 524)

If we accept the admonishments directed at existing theory and research in the previous chapter —for not understanding the role of social and material contexts in shaping environmentally significant behaviours— where might we look for an alternative? There are numerous sociological approaches that try to explain how the social has a role greater than the sum of its parts, that is more than an additional 'factor' or 'driver' of individual behaviour, but instead an intimate part of how we experience the world and our own selves, what we do, how we change and stay the same (Leyshon 2014). In this chapter, the focus is on one perspective in particular, or more accurately a number of loosely affiliated perspectives that share some characteristic features. The perspectives in question are organized around the concept of a 'social practice', so it will be referred

© The Author(s) 2016 **67**
M. Adams, *Ecological Crisis, Sustainability and the Psychosocial Subject*,
DOI 10.1057/978-1-137-35160-9_4

to from here onwards as the 'social practice approach' or 'social practice theory' interchangeably.

Social practice theory has two whole chapters to itself because in the last 10–15 years it is probably the most coherent critical social science alternative to an individualist approach to sustainable behaviour, and the one that has gained the most momentum, particularly in Europe. Evidence of this coherence and momentum can be seen in the inception of a number of research centres, affiliated research grants, and the various public and academic activities associated with them.[1] There are also numerous books, working papers, and journal articles reporting outcomes from this research activity, and debating and developing the approach (e.g. Hand et al. 2005; Sayer 2013; Shove 2003, 2006, 2010a, b, 2012; Shove et al. 2012; Shove and Spurling 2013; Southerton et al. 2004a).[2]

A focus on change lends itself to the study of climate change and sustainability. Sustainable transitions clearly require change. Who or what must change – 'the sustainable development of what?' (Khoo 2013) —is complex and contested, but *that* large-scale social change is required is a tenet of critical social science approaches to the sustainability agenda (Capstick et al. 2014). Social practice approaches are attempts to make sense of social change and stability, to understand the processes involved in how societies endure and alter (Shove et al. 2012, p. 1). Therefore, it is perhaps not surprising then that a significant amount of the development of social practice theory and research in recent years has been dedicated to the issue of climate change, sustainability, and in particular, a critical contribution to behaviour change in this context (e.g. Hargreaves 2011 b; Middlemiss and Parrish 2010; Spurling et al. 2013; Shove 2010a, b, c, d). An exploration of this particular application will follow a broad outline of the concept of social practice.

Social Practices

The first useful thing to say about social practices is that they are not considered an *outcome* of social forces, they are taken 'to be the 'site' of the social' (Shove and Walker 2014, p. 42). This means they are the 'basic unit' of enquiry, when we try to account for change, displacing the explanatory power of, on the one hand, individual variables (attitudes, values, habits), and on the other hand, abstract social systems (technology, economics, politics), as 'barriers' to, or enablers of, change. Making social practices the 'basic unit' of analysis attempts to avoid reducing explanations of the things people do to voluntarism (caused by individuals alone) and determinism (caused by social structures alone). Social practices are doubly generative: they establish the particular possibilities for individual actions, but at the same time 'social life comes into being through practices' (Feldman and Orlikowski 2011, p. 11).

Recwitz's definition of a practice is probably the most cited (e.g. Blewitt 2010, pp. 33–34; Shove and Spurling 2013; Warde 2005, p. 134), though it does not provide a particularly accessible welcome to a newcomer:

> a practice represents a pattern which can be filled out by a multitude of single and often unique actions reproducing the practice ... The single individual – as a bodily and mental agent – then acts as the "carrier"... of a practice – and, in fact, of many different practices which need not be coordinated with one another. Thus, she or he is not only a carrier of patterns of bodily behaviour, but also of certain routinized ways of understanding, knowing how and desiring. These conventionalized "mental" activities of understanding, knowing how and desiring are necessary elements and qualities of a practice in which the single individual participates, not qualities of the individual. (Reckwitz 2002, pp. 249–50)

The key elements of a 'practice' here, to explain this definition further, are that it is a pattern, which is made up of a number of actions, or more precisely 'an organized constellation of actions' (Schatzki 2002, p. 70). The individual—you or I —is not the organizing force *of* these patterns of action (and understanding, knowing how, desiring) however; but rather the vehicle *for* these patterns (routines, conventions), and a meeting place where different patterns come together—'the unique crossing point of practices' (Reckwitz 2002, p. 256). As individuals we take part in reproducing a practice, if we undertake the ways of thinking, feeling and acting that together make it a discernable pattern. In the language of social practice theory, we are in effect 'recruited' to social practices: 'As individuals pass through life, they come into contact with, get recruited to, have 'careers' within, and occasionally defect from a wide variety of different practices' (Hargreaves 2011b, p. 83).

It follows, according to Reckwitz's definition, that the pattern of a social practice is to be found not 'in' the individual who is doing the act (i.e. not organized by their attitudes, values, motives etc.), but in 'elements and qualities' that make a practice recognizable as a practice. In the words of Shove and Walker, both key figures in the contemporary development of a social practice approach, social practices are thus 'recognizable blocks or patterns of activity that are filled out and enacted by practitioners, that is, by those who do, and who, in the enactment and performance of these doings reproduce, transform and perpetuate the practices they carry' (2014, p. 48). In their discussion of household energy demand from a social practice perspective, Shove and Walker draw on Nye's historical account of electricity blackouts (Nye 2010). As an example, it helps illustrate the basic orientation of a social practice approach, and the questions that arise from it, so it is reproduced in full here:

> As he explains, in the 1950s, a power cut would have an impact on office work by affecting lighting and ventilation systems. People might have to go home at dusk but typing and filing would continue as normal. Today a power cut would bring much of what constitutes office activity to a sudden halt. This is just one example and Nye is of the view that large parts of the USA would quickly become uninhabitable if power supplies should fail for any length of time. The point is not just that societies are increasingly dependent on reliable supplies of electricity and oil in particular. For those interested in developing practice-oriented policy… there are two key questions arising from this example: first, how is it that such interconnected bundles and constellations of practices and material arrangements, including technologies of energy provision, distribution and consumption, have taken hold and, second, how might they change? (Shove and Walker 2014, p. 54)

This example nicely indicates how individual actions, in this context, in terms of energy consumption, are understood to be 'recruited' into broader patterns of activity that precede and exceed individual behaviour, and how those social practices can reproduce *and* transform what people do over time. As Shove and Walker make clear here, the key focal points of 'practice-oriented policy' should be an analysis of related practices and the material arrangements that make them possible, with an eye on their vulnerability to change. How exactly this translates into meaningful interventions is less clear, as we shall see, but here we see the basic idea that the social practice is the unit of analysis, not the individual instances of consumption and production that contribute to it.

Elements and Path Dependence

The work of Reckwitz and others develop complex models of practice that take in many components. However, a number of contemporary accounts of a social practice approach (particularly Shove and Pantzar 2005; Shove et al. 2012), while drawing on earlier theoretical developments, make a 'simplifying move' (Shove et al. 2012, p. 23); focusing only on three elements of social practice: materials, competences and meanings. Materials include 'objects, infrastructures, tools, hardware and the body itself' (p. 22); competences consist of 'shared understandings of good or appropriate performance in terms of which specific enactments are judged'; while meanings incorporate 'the social and symbolic significance of participation at any one moment' and include all 'mental activities, emotion and motivational knowledge' (Shove et al. 2012, p. 23).[3] The 'distinctive accumulations' (Shove et al. 2012, p. 146) of these elements are always combined in the performance of a practice: 'Socially acceptable individual behaviour—or the successful performance of a social practice—thus, rests upon the use of objects, tools and infrastructures, of knowledge and skills and of cultural conventions, expectations, and socially shared tastes and meanings. These are the elements that compose social practices' (Røpke 2009, p. 5).

The interdependence of social practices creates a significant degree of 'path dependence' in everyday life. Related terms include descriptions of being 'locked-in' to particularly 'sticky' configurations of practices, pulling in elements of material infrastructure, social meanings and tacit know-how (Shove and Walker 2014; Szerszynski and Urry 2010; Verbong and Geels 2014). This is because of a number of related reasons. Most simply, the number of social practices we can engage in at any one time or over the course of a lifetime is inevitably limited by the demands of time and space. As practices accumulate, they pull in a number of related practices, which come together as 'complexes' and colonize a significant amount of available time, space, and energy (Southerton 2003). Furthermore, the complexes of social practices we do pursue play a hand in shaping what we are likely to undertake subsequently: 'taking one path and not another configures opportunities for the future' (Shove et al. 2012, p. 78). In taking a particular direction we eventually create furrows in which unconnected alternatives, across all elements of practices, become less and less likely. The marshalling of finite material resources behind particular sets of practices makes it likely they are repeated, other practices less so. As Schatzki says (cited in Shove et al. 2012, p. 97) the take up of sets of social practices makes 'courses of action easier, harder, simpler, more complicated, shorter, longer, ill-advised, promising of gain, disruptive, facilitating, obligatory or proscribed, acceptable or unacceptable, more or less relevant, riskier or safer, more or less feasible, more or less likely to induce ridicule or approbation'.

Path dependency is reinforced by the lack of choice we have over many of the practices we engage in. Following Pred (1981) Shove et al. refer to 'dominant projects': 'complexes of practice that orient the ways in which people spend their time and the priorities around which their lives are organized' (Shove et al. 2012, p. 135). Maintaining the 'career' and 'recruitment' metaphor, we might think of dominant projects as complexes of practice we are recruited to via *conscription*. We are obliged to coordinate the elements that enable these complexes—secure resources, make meaningful, develop competence. Institutional roles often fall into this category because they are backed by enforceable rules and regulations, and make the most pressing demands of our time and energy: 'they tend to take time-allocation and scheduling precedence over other projects' (Røpke 2009, p. 2493; Southerton 2003). Institutional roles cited in this context include those related to class, education, ethnicity, families, gender, households, occupation, and broader cultural worldviews and technological systems. In combination, dominant projects significantly 'shape people's sense of what is permissible, desirable and possible' (Szerszynski and Urry 2010, p. 3).

Power and authority are clearly central to the ability of dominant institutions to determine social practices in this way. Consider the role of 'jobseeker' in the UK, and how the interconnected set of enforceable practices captures time and energy in order to receive an income. Rules, laws and expectations, however, are not blanket applications. Who they apply to, and therefore, the specific practices we end up carrying, are in part decided by contingent material factors: 'accidents of birth, history and location' (Shove et al. 2012, p. 66). When and where we are born significantly prefigures the elements we are likely to have access to, and the social practices to which we are recruited. While material resources and infrastructure looms large as an element, they are, for a social practice approach, mutually dependent on meaning and competences to enable social practices and to entrench inequality and injustice (Shove et al. 2012, pp. 135–6).

Accounting for Change: Competition and Collaboration

To sum up what has been said about social practices so far, while they are the 'basic unit' of analysis, they are made possible by combinations of constituent elements. The combination of social practices each individual adopts is significantly shaped by dominant projects. The make-up of dominant projects is determined in the fulfilment of institutional roles that in many regards, we are obliged to undertake. Such pathways reflect and maintain stratified and unequal social arrangements that are difficult to veer off from (Shove et al. 2012, p. 134). It is 'by these means [that] individual lives are woven into the reproduction of societal institutions' (2012, p. 79).

While the metaphor of being woven into societal reproduction through our doings and sayings might appear deterministic, a social practice approach also emphasizes the potential for change. In fact, social change is a history of how complexes of social practices, including dominant projects, are 'made and broken' (Shove et al. 2012, p. 88). These processes are understood in terms of the dynamism of social practices, rather than by reference to individual agency as such. Change is not primarily understood as resulting from an aggregate of attitudes or value changes. Personal agency in the context of a social practice approach, therefore, is an elusive phenomenon, unsurprising when most human activity is framed as performances already prefigured and pre-scripted by existing arrangements of knowhow, meaning and materials.

Shove et al. conceptualize possibilities for change more specifically in terms of 'competition' and 'collaboration' (Shove et al. 2012). Larger collections of social practices are formed by competing or collaborating with other practices in the search for new recruits—remember that time and space constraints limit the number of practices we can be involved it at any one time. Competition is where a practice 'colonises' resources and recruits at the expense of another (Shove et al. 2012, p. 89). They can compete like for like: digital images and slideshows versus material prints and photo-albums, streaming films versus VHS cassettes, sock darning versus cheap clothing; online gaming versus board games; or they can challenge time-use more broadly: gardening versus watching television; shopping versus church attendance. Either way, practices compete over finite time, space and resources in seeking to secure recruits.

Collaboration works differently. It embeds itself into everyday life by making links with existing practices, occupying related practice space or becoming co-dependents on another practice—forming the aforementioned 'complexes' of practices. Personal computer technologies are a good example of collaboration—hitching on to existing practices like exercise, music and navigation to the point that they can appear to be indispensable components of them.

Competition and collaboration are often interwoven across practices and elements—as streaming becomes central to the way we listen to music it melds with the practice of music listening, but it pushes other elements of music listening out of the picture: material (CDs), meaning (the visibility of a CD collection) and competences (CD recording). Unpredictable consequences still ensue—e.g. the vinyl revival. It is not always clear whether a practice has successfully collaborated or competed. Shove et al. (2012) briefly consider the example of family life and the internet. While some commentators decried increasing time spent online as detrimental to family life; others have argued that online developments such as social networking can facilitate and enhance family relationships. Now that the key tenets of a social practice approach have been summarized, we can consider them in the context of sustainability and behaviour change.

Social Practice and Sustainability

> Climate change policymakers and visionaries should hunt down the elements that have the most negative impact upon carbon emissions across a whole group of practices. They should search out and design new elements that would support practices with fewer emissions. Policies would be directed not at bad behaviours, but at 'bad' elements. (cited in Shove et al. 2012, p. 147)   *e.g. cycling infrastructure*

This statement, made anonymously by a climate change workshop participant, is used by Shove et al. to initiate their discussion of how 'certain policy interventions may increase the chances that more rather than less sustainable ways of life persist and thrive' (2012, p. 146). On the surface at least, the implications of a social practice approach for understanding anthropogenic ecological degradation are fairly straightforward. It asserts the importance of context in shaping behaviour, and attempts to define everything that makes up 'context' via the three elements introduced above: materials, competencies and meanings. These three combine to shape human activities in recognizable patterns—including those patterns that cumulatively contribute to climate change and related aspects of the Anthropocene; and those that offer a glimpse of more sustainable alternatives.

The task in hand is to 'identify the practices demanding considerable resources and to study the formation of these practices as a basis for policies' (Røpke 2009, p. 2496). This means that to understand sustainable and unsustainable lifestyles, we need to grasp how the elements come together to make them so—how elements and practices combine to become 'locked-in' or 'sticky'. To understand how a shift to a sustainable society might be orchestrated, we must consider how these combinations might come unstuck, and map the permutations of genuine alternatives.

Translated into intervention, this amounts to a call for a kind of Great Rearrangement: 'transitions of such a scale that conventions, standards, routines, forms of know-how, markets and expectations need rearranging across all domains of daily life' (Shove et al. 2012, p. 163).

This understanding clearly has implications for how we conceptualize both the individual processing of information about sustainability and individual acts of (more or less sustainable) consumption, often considered the rightful targets of sustainability interventions. For a social practice approach most consumption is routine and 'inconspicuous' (Gronow and Warde 2001; Shove and Warde 2002), as Warde spells out accessibly here:

> As individuals we often have limited control over what things we use and how we use them. Convention, infrastructure and shared goals constrain everyone. Types and levels of consumption tend to be determined socially and collectively. A practice-theoretical approach acknowledges this, proposing that consumption is less a matter of individual expression and choice, and more a corollary of the conventions of the range of the specific, socially-organized practices felt to be necessary to live a good life. (Warde 2013, p. 3)

There is an implicit contrast made here with theories of consumption derived from cultural studies that focus 'on consumers as manipulators of symbols engaged in expressing their identity through visible signs, particularly through consumption categories suited to signalling, such as clothes, home decoration, and taste in music and other artistic products' (Røpke 2009, p. 246). In the social practice alternative, our ability to fundamentally alter our own consumption activities is often limited (Brand 2010; Southerton et al. 2004b). This does not mean that consumption is merely a consequence of social systems, rather that both individual consumption and social systems are reflected in the more basic unit of social practices. Like consumption more broadly, the consumption of energy can be understood in this way, as an 'ingredient of social practice' (Shove and Walker 2014, p. 47). The consumption of energy at home, for example, is 'above all an integral part of social practices like living in certain living arrangements, commuting, washing, cooking, eating, driving, etc., all of which are by their very nature systemically – technically, economically, and culturally – interwoven in a particular type of society' (Brand 2010, p. 222). The focus on un/sustainable behaviour, therefore, looks rather different from a social practice perspective.

Informing Interventions

A social practice is only made possible by the particular combination— 'distinctive accumulations'—of elements (Shove et al. 2012, p. 146). It follows that practice-based interventions would begin by asking questions as to how the materials, competences and meanings of a practice 'circulate', and how they come together in ways that make a practice doable; and 'become (or remain) a normal thing to do' (Shove et al. 2012, p. 56). An initial task might, therefore, involve plotting the elements involved in a practice. Here is an extended example provided by Shove et al, using the practice of eating toast for breakfast:

> Rather than imagining one map of practice, we might therefore think of three separate layers– one depicting the distribution of requisite competence, another showing access to necessary materials, and a third plotting the prevalence of toast as a meaningful part of breakfast. Having toast in the morning is only possible where all three layers overlap. In all other situations one or more of the necessary ingredients is missing; for example, toasters are not available, the bread is absent or not of a form that would go in the toaster, or toasters and bread both exist but not the convention of making breakfast at home. The fact that requisite elements co-exist does not guarantee that they will be linked together, but the potential is there. (Shove et al. 2012, p. 45)

If elements are taken to be constitutive of social practices, it makes sense to understand what they are. Advocates of a social practice approach acknowledge that identifying the features key to recruitment, defection and reproduction of 'one practice at a time' (2012, p. 77), especially a fad, is a more straightforward proposition than addressing the complex array of practices involved in something as complex and large scale as anthropogenic ecological degradation and 'sustainable' behaviour. Unsustainable behaviour and lifestyles encompass a range of often tightly related practices, the connections between which we are routinely unaware of. A further level of complexity, if we adhere to a social practice perspective, is found in the dynamic intersection of practices in larger 'bundles' or 'complexes' epitomized by dominant projects (2012, p. 81). Nonetheless, understanding how practices are established, how they develop, spread or die-off, therefore, is possible, and requires, as a starting point, an understanding of the conjunction of the elements involved.

Re-crafting Practices

Those who have developed a social practice approach then focus in on a number of ways in which social practices can change, from relatively moderate interventions to more radical prescriptions for social change.[4] First is a focus on altering the range of elements in circulation, or 'recrafting practices' (Spurling et al. 2013): 'facilitating, or hindering, the availability and circulation of the elements (materials, meanings, competences), of which more and less sustainable practices are formed' (Shove et al. 2012, p. 147). In other words the focus of change here is to squeeze out the elements of a practice that are unsustainable and promote more sustainable alternatives—i.e. less resource-intensive, less wasteful alternatives.[5]

Consider the practice of family meal times. Meal times involve all of the elements that make up a practice. They are meaningful (convenience, health, taste, sociability), involve competence (shopping, storing, cooking, serving), and materials (ingredients, kitchen equipment, energy for freezing, cooking), including the supply chain processes involved 'behind the scenes' (e.g. agricultural practices, transportation, packaging). Spurling et al. ask 'how might the elements of eating practice – ideas of what tastes good, the sociability of eating, and the routine ways in which particular forms of meal are consumed – be re-crafted to direct eating along more sustainable trajectories?' (2013, p. 34). The answer is not, or not only, to change the content of meals, but to focus on the meanings, materials and competences involved in structuring meal times.

An initial task might be to focus on the most resource-intensive aspects of the practice—meat and dairy is commonly the most resource-intensive component of meal times for example (Bailey et al. 2014).[6] A 'recrafting' approach would first aim to understand how elements combine to make meat-eating possible, normal and desirable. Competency might include shopping 'well' for meat (value, quality), as well as storing, cooking and carving it; meanings might include vitality and health, strength; compassion for or dominion over animals; good taste. Materials include the body or body part of the animal, the kitchen equipment, energy for cooling and heating, and the background materials involved in rearing animals, feeding and slaughtering them.

If social practice informed policymakers were to decide that reducing meat consumption was a sustainability target, they would aim to change the elements involved in mealtimes to promote less meat-eating, or change the constitutive elements to make meat-eating less resource-intensive. Taking the former route, interventions might focus on the meanings associated with eating meat (e.g. the role of celebrity chefs and cooking commentators in advocating alternatives to meat); on the material infrastructures that make meat available (e.g. subsidies); and on the skills involved in providing alternatives (e.g. cookery courses, growing your own) (Spurling et al. 2013, p. 35).

What matters for a social practice approach when talking about change is the need to focus on all elements involved in a practice, relating 'to the *meal* and the social context in which it is consumed, rather than the *product* focus of the consumer choice framing' (Spurling et al. 2013, p. 35). In recrafting practices towards sustainability, it is assumed that practices will often persist (family meal times, car driving, heating the home) but that it is possible to shift meanings, competences, materials involved to promote less resource-intense 'versions' of the practice.

Substituting Practices

A second strategy, 'substituting practices', is more radical in that it implies more far-reaching changes in the social fabric (Shove et al. 2012). In the context of sustainable development as social practice approaches tend to frame it, this is about replacing more resource-intensive and/or carbon emitting practices with less intensive ones—new or existing alternatives: 'This framing moves us beyond thinking about the future by extrapolating from existing practices (e.g., personal mobility is heavily car-based therefore a more sustainable transport system will make driving more sustainable) to thinking about how more sustainable practices (new or old) can fulfill the same needs and wants' (Spurling et al. 2013, p. 7). Substituting cycling for driving for example, at least in the UK, would require changes at the level of material infrastructure (e.g. cycle path networks, signage, storage, expanding bike rental schemes), meaning (speed, perceived safety, attached social identities) and competencies (navigating traffic, cycling skills) (Spotswood et al. 2015).

To consider substituting one practice for another depends on an understanding of the elements involved, *and* the ways in which existing practices relate to each other (Shove et al. 2012, p. 152). Promoting participation in carpooling clubs as an alternative to owning a car, for example, 'requires an understanding of the elements that determine car club membership's ability to recruit willing practitioners, as well as understanding how car club membership might be intrinsically connected to other practices (e.g., doing a large, weekly grocery shop)' (Kurz et al. 2015, p. 122).

Interlocking Practices

Thinking about the relationship between practices might be the basis for substituting one for another, but it also informs a still more ambitious intervention – changing how practices 'interlock' (Spurling et al. 2013). This third strategy reflects the tenet of social practice approach discussed above, the idea that complexes of practices end up tightly associated, carrying out one practice firmly implicating us in carrying out others (physical exercise, mobility, diet). The more we are obliged, or feel obliged, to undertake one practice in relation to another (e.g. to drive to the gym), the 'stickier' the complex. It follows then that the shift to less resource-intensive practices needs to address the enduring links between them—the 'circuits of reproduction' (Shove et al. 2012, p. 16).

Mobility, to take a prime example, has become increasingly strongly linked to car driving, which as a practice has become embedded in how and where we live, shop, work and holiday (Watson 2012). As Shove et al. describe it, 'the petrol and steel car has been systematically locked into the organization of society. Cars have become progressively embedded through patterns of economic and suburban development and through the remaking of space and time in ways that demand and assume a relentless logic of automobility' (Shove et al. 2012, p. 154). Intervention at the level of 'circuits of reproduction' involves a radical rethink of how practices intersect to create the need for mobility. In other words, when we are recruited to private car use, what other practices are we signing up to? We might analyse what gets done via the car in terms of meanings, materials and competences: such as work, childcare, shopping and leisure activity. Broader structural practices then come into play, such as catchment areas for schools and related policies—who drives to school and why? This is an enormous challenge, demanding, in short, the reconfiguration of 'the spatial arrangement of everyday practices' (Spurling and McMeekin 2015, p. 88). More specifically, it 'takes us away from road or even transport policy as the only context for interventions. Broader aspects of urban planning are clearly relevant; policies to support ever more efficient communications network might serve as a partial substitute for the transport network' (Spurling et al. 2013, p. 31). At a more localized level, this translates into how we experience the taken-for-granted and routine ways in which we go on in everyday life, as it is embodied in our habits, and embedded in material and symbolic social arrangements. Change at this level is about the possibility of 'recalibrating' personal and collective normality in ways that reproduce sustainable ways of life (Shove et al. 2012, p. 157).

The role for any government seeking a radical shift towards sustainability would be, on the back of this understanding, to convene those involved in steering dominant projects and institutional roles, 'bringing existing actors together (i.e. businesses, manufacturers, marketing organizations, retail outlets) as part of a deliberate strategy to reconfigure the character and the distribution of the elements of which more sustainable practices *could be* made, and in seeking to break the ties that hold other less sustainable arrangements in place' (Kurz et al. 2015, p. 122). Such hopes sidestep the extent to which such a steering role might challenge or conflict with existing government agendas and policies, and the broader context of neoliberal consumer societies which they reflect. Such tension may actually manifest as the active suppression or marginalization of alternatives rather than their embrace. At the very least, a fuller analysis of 'regime resistance' (Geels 2014), and related critical analysis of power is required in addition (Sayer 2013); a point developed in the next chapter.

Relatedly, for a social practice approach explicitly advocating radical change (Capstick et al. 2015), a potential seam of inquiry is the 'social forms' that might 'hasten the adoption of more rather than less sustainable practices' (Shove et al. 2012, p. 160). Here we see a potential connection with 'prefigurative' and 'alter'-politics (Hage 2015), and with the practice and study of intentional sustainable communities (Raco 2005). Although not necessarily drawing on a social practice approach, 'communities of practice' dedicated to the creation of alternative social forms in effect strive to reconfigure everyday life in ways that make possible the simultaneous and interconnected pursuit of sustainable practices. However partial and imperfect, they prefigure the 'sorts of bonds and links [that] might emerge from, and enable, the recurrent enactment of lower impact ways of life' (Shove et al. 2012, p. 160). Again, work more explicitly addressing such possibilities is considered in subsequent chapters.

To summarize, as interventions directed at encouraging sustainability, a social practice approach has focused on reducing the resource-intensity and carbon emissions of social practices. The various strategies devised reflect a continuum, from minor adaptation to more radical upheaval. In 'recrafting' the elements involved in social practice, (e.g. tax-breaks on more fuel-efficient cars, improving the efficiency of driving), there is some overlap with existing behaviour change strategies (Capstick et al. 2015). However, in acknowledging the links between practices, *substituting* practices places more emphasis on the social and material context in which driving or cycling becomes normalized; and therefore, on context as a target for intervention. In focussing on how practices *interlock* there is a greater and more explicit emphasis on intervening across a number of practices.

In this chapter, I have considered how sustainable behaviour has been theorized within the recent 'practice turn'. Social practice theory offers an impressive array of conceptual tools within a broader framework, building subtly on a range of perspectives. In placing social practices centre stage, recognition of the need to urgently understand the role of social context in helping or hindering sustainable development. Beyond that recognition, there is a great deal of complexity offered in discerning how social practices are understood, and how they contribute to change and stasis on a societal level—intersecting elements, cross-referenced and monitored across many practices; social practices in competition and collaboration; forming bundles, complexes and dominant projects.

The inclusion of the material as an element is perhaps the most potent contribution of the social practice approach. It anchors the accounts of social conventions, meanings and embodied knowhow to the materiality of space, time and resources. The desire to forefront the dynamism of social practices is also vital. From a sociological and historical perspective, the contingent dominance of some social practices over others in particular times and places is made more visible, empirically and theoretically. As a result, the potential for change, even radical change, in the complexes of social practices that maintain high-carbon, unsustainable ways of life is made to seem challenging, fraught and uncertain, but potentially achievable nonetheless. However, this does not mean a complete picture of propensities for sustainable or unsustainable behaviour has emerged, in which all the intersections of social, natural, cultural, political and psychological are successfully enfolded. In the next chapter, three overlapping critical issues are tackled—meaning, nature and power.

Notes

1. Research groups include the Sustainable Practices Research Group (SPRG), 2010–14 and the Dynamics of Energy, Mobility and Demand (DEMAND) Centre (2013–18). SPRG was a collaboration between various universities and funded by the ESRC as well as the Scottish Government and DEFRA. 'The broad aim of the project is to enhance the social scientific understanding of how habits in areas of everyday consumption form and reproduce'. More information is available at http://www.sprg.ac.uk. The DEMAND Centre is another collaborate project, also funded by the ESRC (as well as Transport For London and the International Energy Agency). It 'takes a distinctive approach to end use energy demand, recognizing that energy is not

used for its own sake but as part of accomplishing social practices at home, at work and in moving around'. More information is available at http://www.demand.ac.uk

2. It is beyond the scope of this chapter to convey the intricacies of the diverse social practice literature that has sprung up to better understand 'sustainable behaviour change'. The intention is to capture some of the distinctive common features that tend to be shared by the various retellings of a social practice approach. My emphasis is on features I consider relevant and useful in approaching sustainability, so nuances will inevitably be missed. Neither is there space here to dwell on the theoretical antecedents and foundations of a social practice approach or a supposed 'practice turn' (Schatzki et al. 2001; Shove and Walker 2014, p. 46). Social practice advocates state influences including Giddens, Bourdieu, Schatzki and Reckwitz amongst others. However, the focus here is on the contemporary take up and application of social practice approaches to the sustainability agenda, so there will be little discussion of those theoretical origins, unless they bear a direct relevance to the point in hand.

3. In a 'short-circuiting' of problems around how meaning relates to subjectivity and agency, the authors 'treat meaning as an element of practice, not something that stands outside or that figures as a motivating or driving force' (p. 23).

4. Spurling et al. (2013) loosely convert Shove et al.'s more abstract emphasis on four dimensions of how social practices change (2012) into three strategies for interventions. The summary here uses Spurling et al.'s categories but imports some of the detail from Shove et al. and others.

5. In the detail of the many examples given across the social practice literature, there is ambiguity about what counts as elements of practices, practices in their own right, and practices as part of larger bundles and complexes – is long-distance flight an element of other practices (conference attendance, tourism), a practice in its own right made up of various elements (navigating airports, access to finances) or a practice that is part of a complex (globalized family life)? It is of course possible that what is a practice in one situation can be an element in another, but the extent of this cross-referencing can make the conceptual categorization so elastic that it can at times feel redundant.

6. Capstick et al. (2014, p. 3) also cite Bailey et al. in claiming that 'meat and dairy products alone represent a great share of [carbon] emissions than those deriving from all worldwide road transportation, trains, shipping and air travel'.

References

Bailey, R., Froggatt, A., & Wellesley, L. (2014). *Livestock–climate change's forgotten sector*. London: Chatham House.

Blewitt, J. (2010). *The ecology of learning: Sustainability, lifelong learning and everyday life*. London: Routledge.

Brand, K. W. (2010). Social practices and sustainable consumption: Benefits and limitations of a new theoretical approach. In M. Gross & H. Heinrichs (Eds.), *Environmental sociology: European perspectives and interdisciplinary challenges* (pp. 217–235). Netherlands: Springer.

Capstick, S., Lorenzoni, I., Corner, A., & Whitmarsh, L. (2014). Prospects for radical emissions reduction through behavior and lifestyle change. *Carbon Management, 5*(4), 429–445.

Capstick, S., Whitmarsh, L., Poortinga, W., Pidgeon, N., & Upham, P. (2015). International trends in public perceptions of climate change over the past quarter century. *Wiley Interdisciplinary Reviews: Climate Change, 6*(1), 35–61.

Feldman, M. S., & Orlikowski, W. J. (2011). Theorizing practice and practicing theory. *Organization Science, 22*(5), 1240–1253.

Geels, F. W. (2014). Regime resistance against low-carbon transitions: Introducing politics and power into the multi-level perspective. *Theory, Culture and Society, 31*(5), 21–40

Gronow, J., & Warde, A. (Eds.). (2001). *Ordinary consumption*. Hove: Psychology Press.

Hage, G. (2015). *Alter-politics: Critical anthropology and the radical imagination*. Melbourne: Melbourne University Publishing.

Hand, M., Shove, E., & Southerton, D. (2005). Explaining showering: A discussion of the material, conventional, and temporal dimensions of practice. *Sociological Research Online, 10*(2). http://www.socresonline.org.uk/10/2/hand.html. Accessed 18 Dec 2015.

Hargreaves, T. (2011b). Practice-ing behaviour change: Applying social practice theory to pro-environmental behaviour change. *Journal of Consumer Culture, 11*(1), 79–99.

Khoo, S. (2013). Sustainable development of what? In F. Fahy & H. Rau (Eds.), *Methods of sustainability research in the social sciences*. London: Sage.

Kurz, T., Gardner, B., Verplanken, B., & Abraham, C. (2015). Habitual behaviors or patterns of practice? Explaining and changing repetitive climate relevant actions. *Wiley Interdisciplinary Reviews: Climate Change, 6*(1), 113–128.

Leyshon, C. (2014). Critical issues in social science climate change research. *Contemporary Social Science: Journal of the Academy of Social Sciences, 9*(4), 359–373.

Middlemiss, L. K., & Parrish, B. (2010). Building capacity for low-carbon communities: The role of grassroots initiatives. *Energy Policy, 38*(12), 7559–7566.

Nye, D. E. (2010). *When the lights went out: A history of blackouts in America.* Cambridge, MA: MIT Press.

Pred, A. (1981). Social reproduction and the time-geography of everyday life. *Geografiska Annaler. Series B, Human Geography, 63*(1), 5–22.

Raco, M. (2005). Sustainable development, rolled-out neoliberalism and sustainable communities. *Antipode, 37*(2), 324–347.

Reckwitz, A. (2002). Toward a theory of social practices. A development in culturalist theorizing. *European Journal of Social Theory, 5*, 243–263.

Røpke, I. (2009). Theories of practice – New inspiration for ecological economic studies on consumption. *Ecological Economics, 68*(10), 2490–2497.

Sayer, A. (2013). Power, sustainability and well-being: An outsider's view. In E. Shove & N. Spurling (Eds.), *Sustainable practices: Social theory and climate change* (pp. 292–317). London: Routledge.

Schatzki, T. R. (2002). *The site of the social. A philosophical account of the constitution of social life and change.* Pennsylvania: The Pennsylvania State University Press.

Schatzki, T. R., Knorr-Cetina, K., & Von Savigny, E. (2001). *The practice turn in contemporary theory.* Hove: Psychology Press.

Schwanen, T., Banister, D., & Anable, J. (2012). Rethinking habits and their role in behaviour change: The case of low-carbon mobility. *Journal of Transport Geography, 24*, 522–533.

Shove, E. A. (2006). Efficiency and consumption: Technology and practice. In T. Jackson (Ed.), *The Earthscan reader in sustainable consumption* (pp. 293–305). London: Earthscan.

Shove, E., & Pantzar, M. (2005). Consumers, producers and practices: Understanding the invention and reinvention of Nordic walking. *Journal of Consumer Culture, 5*(1), 43–64.

Shove, E. (2010a). Sociology in a changing climate. *Sociological Research Online, 15*(3), 12 http://www.socresonline.org.uk/15/3/12.html

Shove, E. A. (2010b). Beyond the ABC: Climate change policy and theories of social change. *Environment and Planning. A, 42*(6), 1273–1285.

Shove, E. A. (2010c). Social theory and climate change questions often, sometimes and not yet asked. *Theory, Culture and Society, 27*(2–3), 277–288.

Shove, E. A. (2010d). Submission to the House of Lords Science and Technology. Select Committee Call for Evidence on Behaviour Change.

Available at: http://www.lancs.ac.uk/staff/shove/transitionsinpractice/papers/ShoveHouseofLords.pdf

Shove, E. A. (2012). Habits and their ceatures. In A. Warde & D. Southerton (Eds.), *The habits of consumption* (pp. 100–113). Helsinki: Helsinki Collegium for Advanced Studies.

Shove, E. A., & Spurling, E. A. (2013). Sustainable practices: Social theory and climate change. In E. A. Shove & N. Spurling (Eds.), *Sustainable practices: Social theory and climate change* (pp. 1–15). London/New York: Routledge.

Shove, E., & Walker, G. (2014). What is energy for? Social practice and energy demand. *Theory, Culture & Society, 31*(5), 41–58.

Shove, E. A., & Warde, A. (2002). Inconspicuous consumption: The sociology of consumption, lifestyles and the environment. In R. Dunlap, E. Buttel, P. Dickens, & A. Gijswijt (Eds.), *Sociological theory and the environment: Classical foundations, contemporary insights* (pp. 230–251). Lanham: Rowman and Littlefield.

Shove, E. A., Pantzar, M., & Watson, M. (2012). *The dynamics of social practice: Everyday life and how it changes*. London: Sage.

Southerton, D. (2003). 'Squeezing time' – Allocating practices, coordinating networks and scheduling society. *Time and Society, 12*(1), 5–25.

Southerton, D., Chappells, H., & van Vliet, B. (Eds.). (2004a). *Sustainable consumption: The implications of changing infrastructures of provision*. Cheltenham: Edward Elgar.

Southerton, D., Warde, A., & Hand, M. (2004b). The limited autonomy of the consumer: Implications for sustainable consumption. In D. Southerton, H. Chappells, & B. Van Vliet (Eds.), *Sustainable consumption: The implications of changing infrastructures of provision* (pp. 32–48). Cheltenham: Edward Elgar.

Spotswood, F., Chatterton, T., Tapp, A., & Williams, D. (2015). Analysing cycling as a social practice: An empirical grounding for behaviour change. *Transportation Research Part F: Traffic Psychology and Behaviour, 29*, 22–33.

Spurling, N., & McMeekin, A. (2015). Interventions in. In N. practice: Sustainable mobility policies in England. In Y. Strengers & C. Maller (Eds.), *Social practices, intervention and sustainability: Beyond behaviour change*. London: Routedge.

Spurling, N., McMeekin, A., Shove, E., Southerton, D., Welch, D. (2013). *Interventions in practice – Re-framing policy approaches to consumer behaviour*. Manchester: Sustainable Practices Research Group. http://www.sprg.ac.uk/uploads/sprg-report-sept-2013.pdf. Accessed 12 Sep 2016.

Urry, J. (2010). Sociology facing climate change. *Sociological Research Online, 15*(3), 1.

Warde, A. (2005). Consumption and theories of practice. *Journal of Consumer Culture, 5*(2), 131–153.

Warde, A. (2013). Sustainable consumption and behaviour change. *Discover Society, 1.* http://www.discoversociety.org/2013/10/01/sustainable-consumption-and-behaviour-change/. Accessed 18 Dec 2015.

Watson, M. (2012). How theories of practice can inform transition to a decarbonised transport system. *Journal of Transport Geography, 24,* 488–496.

5

Power, Nature and Meaning: Critiquing a Social Practice Approach to Sustainability

Introduction

In Chapter 3 a number of criticisms of psychological approaches to ecological crisis and behaviour change were considered. A useful way to initiate a critical discussion of the value of a social practice approach is to reflect on how well it addresses these criticisms in offering an alternative. The six critical points were as follows: Underestimating the nature and scope of change required; depoliticizing the ecological crisis; ignoring the power of conflicting interests; reifying citizens as passive subjects; fixing behaviour in stasis; and neglecting the importance of social context. Each will now be briefly considered in turn. First, a social practice extends the scope of change required considerably, inviting not just a fuller account of the social, but one in which material infrastructure and social relations intersect with embodied ways of navigating the world. This clearly translates into the call for more ambitious programmes of change across the critical social sciences (e.g. Capstick et al. 2015). Questions remain over the extent to which social practice theory appreciates the fundamental ways in which human social practices are embedded in nonhuman nature,

© The Author(s) 2016   **89**
M. Adams, *Ecological Crisis, Sustainability and the Psychosocial Subject*,
DOI 10.1057/978-1-137-35160-9_5

and how change might depend on transformations in social practices at this fundamental level of interrelatedness. 'Nature' or 'crisis' makes few appearances in social practice accounts of social change—other than as an inert 'material' and resource, which we can use more or less sustainably. This point is addressed below as one of three areas for critique.

Second, a social practice approach politicizes anthropogenic ecological crisis in that governments and government policy are designated a central role in configuring, and incentivising, 'dominant projects'. These projects amalgamate practices into 'complexes' that significantly eat into available time and resources, and shape the elements that are needed to make up practices. It also has far-reaching political-ideological implications for the sustainability agenda in that it divests it of the trenchant individualism that marks mainstream policy approaches to behaviour change. Social practice theory encourages a welcome shift to focus on society-level transformation. There is less development of the extent to which the meaning of crisis and appropriate responses are culturally and personally derived from, or feed into political interests; or of the power dynamics involved in hindering and opposing the level of change suggested. The issue of power is taken up below.

The third point can be dispensed with quickly—there is undoubtedly a recognition that practices are dependent on interlocking elements, and that environmentally significant behaviours or 'performances' are only likely to change if relevant aspects of those elements are modified or transformed. The fourth and fifth points can be covered together. There is undoubtedly a great deal of dynamism imported into the ways human activity is implicated in social practices. The skilled, habitual and embodied nature of everyday competencies are central to the dynamics of social practice, and therefore, to change. However, our role as active negotiators of the 'problem' of anthropogenic ecological crisis is given less emphasis in a social practice approach. There is very little said about the ways in which we actively contest, negotiate and make sense of the reality (or unreality) of anthropogenic ecological degradation, including how we feel and talk about 'it' (or studiously avoid feeling or talking about it). We can only assume that this is not considered significant in shaping what we say and do in relation to anthropogenic ecological degradation. Here,

we address the issue of meaning. In terms of the sixth concern, social practice approaches undoubtedly incorporate a much richer portrayal of social context, but do they begin to grasp the psychosocial satisfactorily? Addressing these three critical points, relating to power, nature and meaning, is the basis for an answer.

① **Power**

[handwritten annotations: Which forms of power can religion address/inhibit? — direct ✓ — discursive ✓ (framing) = material ✗ (minimal) — institutional? not in UK]

The first of our criticisms relates to a 'missing account' of power in relation to social change and stasis (Schönian and Laube 2013, p. 126; Maniates 2014). While they offer the most detailed articulation of social practice approach as a basis for understanding un/sustainable behaviour, Shove et al. acknowledge that they 'do not go deep into questions of power' (2012, p. 120). There are promising signs of acknowledgment in passing, such as the comment that 'seemingly neutral circuits of reproduction are skewed and slanted by patterns of inequality, these being patterns that are in turn perpetuated through the dominance and marginalization of specific practices and practice complexes' (2012, p. 117); but details never fully emerge. The language of a social practice approach arguably militates against a serious consideration of power and the contestation of power. While individuals are 'carriers' of practice, are 'recruited' to social practices, and might 'defect' for example, their reluctance or otherwise is rarely considered. Although differentiation between more or less powerful practitioners is described as significant, a distinction is only made in terms of the ability of the more powerful to define the parameters of a practice, not how and why they may contribute to anthropogenic ecological degradation disproportionately.

Sayer calls for the import of political economy analysis into the social practice approach to sustainability and climate change, specifically to address what he considers to be the neglected issue of power (Sayer 2013). Geels' analysis of 'regime resistance towards low-carbon transitions' offers an advance on this front (Geels 2014; see also Turnheim and Geels 2013). His critical foil is not social practice approaches as such, but the 'multi-level perspective' (MLP) on transitions to sustainability (e.g. Geels 2002, 2012; Smith et al. 2005; Geels and Schot 2007). However, there are many parallels between the way MLP and social practice approaches frame sustainable behaviour (Capstick et al. 2015). Briefly stated, MLP recognizes that addressing anthropogenic ecological degradation requires an understanding of 'sociotechnical transitions' across various levels that include 'green' technologies… new infrastructures, user practices, policies and cultural meanings' (Geels 2014, p. 23).

More importantly, Geels identifies similar shortcomings, and the subsequent analysis of power provided to address them is directly relevant to a social practice approach.

Geels's primary focus in terms of inaction is not reluctant or ambivalent citizen-consumers (though these are symptoms of the situation described), or the bundles of social practices they are caught up in that inadvertently reproduce unsustainable energy demand, consumption or mobility but 'active resistance by incumbent regime actors' to fundamental change (2014, p. 22). Power and resistance to change is conceptualized in terms of 'the ways in which these actors use power and politics to resist fundamental transitions to new low-carbon systems' (2014, p. 23). Geels combines existing accounts of power in relation to the sustainability agenda to provide a 'dynamics' of the power to resist fundamental change (e.g. Levy and Newell 2002; Avelino and Rotmans 2009; Grin 2010; Kern 2011). He categorizes the 'dynamics' as involving instrumental, discursive, material and institutional forms of power and resistance. He explores them in the context of the UK electricity system, against the backdrop of the recent renaissance of nuclear and natural gas. Electricity generation significantly contributes to carbon dioxide emissions in the UK, and despite ambitious climate change targets by successive governments, reduction in carbon emission from the electricity sector, and transition to renewable forms of energy, has been slow. This is in part because of the continuing use of gas, coal and nuclear energy which in turn, argues Geels, reflects the 'resilience and resistance of incumbent regimes' (2014, p. 23). The threat posed by the entrenchment and opposition of existing fossil fuel regimes is most explicit in recent fossil fuel divestment movement and 'Keep it in the Ground' campaign.[1] Continuing to burn the world's 'reserves' of carbon will undoubtedly means exceeding the conservative 2° C target set for restricting climate change (Berners-Lee and Clark 2013).

The use of instrumental power is the most straightforward and 'direct': it refers to when actors use available resources to pursue goals and interests in immediate interactions—a combination of material and relational power. Geels' examples include 'positions of authority, money, access to media, personnel, capabilities' (2014, p. 28). In the example of the UK Electricity, Geels described the then prime minister Tony Blair going ahead with a nuclear agenda regardless of the outcome of public consultation, and the decision to subsidize Electricite de France's supply of nuclear energy despite public promises not to do so.

Geels' second form of power is discursive, referring to attempts to 'shape not only what is being discussed (thus setting agendas) but also how issues are discussed' (2014, p. 29). It reflects the importance attributed to the 'framing' of environmentalist issues, particularly by powerful groups, elsewhere in the social sciences (e.g. Dunlap and McCright 2010, 2011; Hoffman 2011; Lakoff 2010; Randall 2009; Wilk 2010). The utilization of instrumental (material and relational) power obviously helps here, e.g. ownership of newspapers, funding of 'think tanks', access to media commentators, shared social networks (Jacques et al. 2008). Geels delineates how discursive power is important at multiple levels of framing; how problems are identified and defined (diagnostic); which solutions are advanced (prognostic); and the development of rationales intended to drive and inspire action (motivational) (2014, pp. 29–31).

In terms of the UK electricity sector Geels describes how the 'problem' of climate change and its relation to energy production has become established, through discourses around solutions that have actively managed support for incumbent regimes of coal, nuclear and gas production. This has been at the expense of civil society alternatives such as the Transition Town movement or community energy projects. We might add more recent attempts to promote the hydraulic fracturing (fracking) of shale rock to release natural gas as a solution to the problem of climate change (Klein 2014). In this context, motivation framing addresses attempts to dampen the sense of urgency around the need to transition to a sustainable society, and the scale of change that might be required; and to emphasize instead, the value of cheap energy or energy security (Geels 2014, p. 33).

Material strategies incorporate the use of financial and technological capital to fund and support technological development, secure funding, and influence regulation. The strategy here is to improve the 'efficiency' of current 'socio-technical regimes', or make promises (utilizing discursive strategies) about future improvements, and therefore, maintain their stability. There are many parallels here with the analysis of 'techno-fix solutions' to ecological crisis that play to existing faith in human mastery over nature and reluctance to counter deeper change (Randall 2009). In

Merchants of Doubt.

the context of electricity, technical improvements to the emissions associated with the burning of coal such as carbon-capture storage (CCS), supports the circulation of 'clean coal' discourses. More recently, hopes and promises have been extended to the ambitious geo-engineering of the climate, which maintains a focus on existing regimens of technology and finance (Klein 2014).

Geels' fourth and final form of power is institutional power, which is 'embedded in political cultures, ideology and governance structures' and further facilitates the strategies of incumbent actors described above (2014, p. 34). Examples he notes include the liberal market economy, through which minimal state regulation, pro-business sympathies and faith in 'the market' is legitimized. It supports a technocratic and cost-benefit analysis approach to policy-making, and frames 'low-carbon transitions as a techno-economic management challenge' (p. 34). The promotion of a 'post-political' discourse masks the contested nature of social and political responses to anthropogenic ecological degradation policy-making, and plays down the scope for agency, by presenting the solutions that suit incumbent actors as the only reasonable options, e.g., 'there is no alternative to nuclear power' (Geels 2014, p. 35). Behind the post-political mask, however, the process of policymaking remains firmly political, embedded in what he describes as:

> informal consultation networks that give industry groups privileged positions as providers of technical knowledge and advice, with limited access for outsiders to close-knit policy networks. This policy style has made UK policymakers more skilful in dealing with incumbent firms and technical experts than with citizens, cities, and social movements, which helps explain why large-scale technical options receive more attention and funding than alternative transition pathways. (Geels 2014, p. 34)

Geels' analysis, in sum, elucidates how the forms and strategies of power intersect dynamically to create and consolidate a 'core alliance' between policymakers and incumbent firms, through which 'regime stability' is preserved (2014, p. 26). The power of this alliance to resist change is multiplied by the many shared social networks and frequent contact between senior policymakers and big business. The way decisions about policymaking, public contracts, research funding etc. are embedded in the shared and overlapping networks of privileged senior politicians, lobbyists, newspaper owners, public school ties and corporate power is a recurring topic of investigative journalism and academic work (e.g. Khan 2011; 2012a, b; Klein 2014).[2] The question of how this alliance creates broader stability and legitimacy in relation to sustainability in society is addressed less convincingly. For Klein, money and access serves the narrow self-preservation instinct of the network, which 'gets in the way' of 'our collective self-preservation instinct' (2014, p. 149). She elaborates briefly:

> Environmentalists often speak about contemporary humanity as the proverbial frog in a pot of boiling water, too accustomed to the gradual increases in heat to jump to safety. But the truth is that humanity has tried to jump quiet a few times. In Rio in 1992. In Kyoto in 1997. In 2006 and 2007, when global concern rose yet again after the release of *An Inconvenient Truth*, and with the awarding of the Nobel peace Prize to Al Gore and the Intergovernmental Panel on Climate Change. In 2009, in the lead up to the United Nations climate summit in Copenhagen. The problem is that the money that perverts the political process as a kind of lid, intercepting that survival instinct and keeping us in the pot (2014, pp. 149–50)

Despite Klein's later emphasis on the viability of a ground-up climate action movement, most of her examples of attempting to jump out of the pot involve top-down events. Geels, adopting Levy and Newell's 'neo-Gramscian political economy', argues that a stable hegemony requires 'consensual legitimacy in civil society', which is created and maintained by 'widely accepted discourses' (2014, p. 27). This seems to unnecessarily restrict his earlier analysis of power to the operation of discourse, and to exaggerate the extent to which active consent and legitimacy is required as a component of regime stability. As Sayer argues, the micro-politics of everyday life may well involve resistance, contestation and reluctant compliance, rather than acceptance, even when we follow routine situations habitually (Sayer 2013, pp. 168–9). In the context of anthropogenic ecological degradation, the forms of power Geels identifies, as he acknowledges, may blunt, obscure, marginalize and belittle contestation and alternative pathways, but not necessarily eradicate them.[3]

Shove and Walker argue that an emphasis on power and politics in accounting for a change in the context of low-carbon transitions tends to unhelpfully rely on 'abstracting sets of forces or systems' (2014, pp. 45–7). Geels' analysis and related accounts of how power and privilege are maintained are insightful in that they engage with the detail of how power might be reproduced in social practices, and in doing so provide a dynamic understanding of active resistance by incumbent actors. This approach is broadly amenable to viewing 'regime stability' as a complex of social practices—it involves a dynamic intersection of different competencies, meanings and materials. Thus Shove and Walker's reminder of what a social practice focus should be, set out in contrast to abstraction, arguably applies to his analysis: 'detailing precisely how social practices, and bundles and constellations of practice, hang together, and of identifying the material and other arrangements amidst which they 'transpire', and which they also sustain and reproduce' (Shove and Walker 2014, p. 47).

If the social practices involved in 'regime stability' are a vital ingredient in Geels's analysis of (not) transitioning to low-carbon societies, what does their absence from a social practice approach tell us? In emphasizing meso-level domain of social practices and everyday life, a particular vision of that 'everyday life' is reified as the site of enquiry. Although elements are mapped that exceed their everyday setting, the endpoint of analysis is still how and where they 'hang together' in an imagined everyday of consumer-citizens, e.g., the 'dynamics of demand' (Shove and Walker 2014, p. 52). In this framework, electricity demand is 'a normal and necessary part of doing things like lighting, cooking and heating' (2014, p. 49). Geels points us to more complex constellations of practice in settings that are no less 'everyday' for some, involve more than demand dynamics, and which are central to an analysis of an issue such as the production and consumption of energy. This sequestration of the 'everyday' as a routine, inconspicuous vehicle for accomplishing social practices evacuates it of politics and power, resistance and contestation, which are essential at micro and macro levels.

They also point the way to additional concerns for a critical research agenda. The practices of regime resistance, involving instrumental, material, discursive and institutional power (including violence and imprisonment) need to be firmly kept in focus as part of political economic analysis, otherwise, a social practice approach 'might be said to accommodate and merely qualify, rather than challenge, the tendency to load all the responsibility onto individuals' (Sayer 2013, p. 176). One possibility is to better understand the dynamics of power and privilege (e.g. Keister 2014; Khan 2011; Sayer 2014); including micro-level 'performances' of practices amongst 'high-carbon lifestyles' (Capstick et al. 2015) and elite groups (e.g. Khan and Jerolmack 2013); and research trained more specifically on the social practices involved in 'regime resistance'. Geels' analysis also suggests that researching how existing regimes central to resisting change might be destabilized and dismantled (Turnheim and Geels 2013) might also be of great value; including scholarship targeting the 'ruse of elite rhetoric' in this context (Khan 2012b, p. 482). A logical conclusion in light of Geels' analysis is that 'socio-political struggles' and 'politically inspired regime destabilization' involving fossil fuel companies and related business is necessary, challenging the core alliance and the conduits of their power to maintain stability. A research agenda which prioritizes these issues will contribute to Capstick et al.'s demands for a critical social science of sustainability, one that offers more 'forthright advocacy' of radical change (2015, p. 10).

Issues of how to encourage collective mobilization and where it might be most effective remain pertinent, alongside the struggle to develop alternatives. Accordingly, individual (in)action, though conceptualized differently, is still an important focus, for the possibility of an adequate response to anthropogenic ecological degradation. 'Socio-political struggles' and 'regime destabilization' rely on the potential for contesting the forms of power identified by Geels and others. This engages us with the micro-politics of everyday life and the contestation of meaning, discussed below, as a potentially vital ingredient of everyday life, but one that is marginalized by a social practice approach.

Nature

Nonhuman nature is arguably another missing 'element' from social practice theory, in general terms and more specifically as one that constitutes social practices; or perhaps more accurately, the relationship between human beings and the rest of nature. Of course, 'nature' is a social construct: the meaning and value of 'nature' is historically, culturally and socially contingent. The routine ways in which we understand and experience 'nature' are bound up in the structure of everyday life, organized, as we have seen, by bundles and complexes of social practices which we embody as competences and in which we are materially and socially embedded. Accepting this premise in a critical realist spirit it is nonetheless incumbent upon any detailed engagement with sustainability to make some kind of sense of what is at stake—of what is to be sustained (Khoo 2013).

Ironically, the application of a social practice to sustainability has almost nothing to say about the relationship between humans and the rest of nature. In fact, for the most part, it reproduces a culturally contingent discourse of nature as an external, inert, pliable resource. It follows that for 'unsustainable behaviour' read 'resource-intensive' and/or 'high carbon emitting' practices—driving, heating and meat-eating. However, the meanings of 'unsustainability' do not just relate to efficiency and pollution. Social practice approaches are in danger of reifying a 'dissociated rationalism' in making it so (Rustin 2013). They are contributing to a skewed techno-scientific discourse, in which the ecological crisis is an engineering problem—albeit social as well as technological. Social practices are consequently presented as harmful in that they are resource-intensive—but otherwise benign; removing or at least marginalizing the possibility of any ethical, normative or political critique of practices (Cudworth 2015).

Social practice theories relating to sustainability also neglect how 'circuits of reproduction' facilitate or obfuscate social practices that permit connection to nonhuman nature—places, species, and related forms of attachment and care (Frantz and Mayer 2014; Vining 2003). Primarily, this is because such practices do not enter the 'range of convenience' for social practice theory, with a sociological heritage long suspicious of 'essentialist' concepts of nature and the natural world (Crist 2004). Even so, it seems extraordinary that social practice approaches ostensibly dedicated to sustainability issues make little conceptual space for the possibility that the ecological crisis itself, and our relationship to nonhuman nature, in some ways impact upon what we feel, think, say and do; and that this impact might be significant in mobilizing us to act differently (Fisher 2013a, b). After all this might be precisely, to revisit Soron, 'where the agency and cultural will for such change might reside' (Soron 2010, p. 179).

Perhaps social practice theory represents a broader difficulty in articulating the human-nonhuman nature relationship—a relationship that is routinely discursively unfamiliar and unavailable in contemporary capitalist,

which could
include
religious practices

industrialized societies (Franklin 1999). Nonetheless, there are attempts to do so, across a range of developing accounts in the humanities and the social sciences. Fisher, for example, refers to 'recollective practices' as 'those activities that aim more directly at recalling how our human psyches are embedded in and nurtured by the larger psyche of nature and at learning the essentially human art of revering, giving back to, and maintaining reciprocal relations with an animate natural world' (2013a, p. 13). Although Fisher uses the term 'practices' here, his description is clearly articulated in a very different register from that of social practice approaches; one that is developed with growing subtlety across a number of disciplines. A number of these developments, and how they might be incorporated into a psychosocial approach, are explored in more detail in Chapter 10.

Meaning

Shove et al. acknowledge that their model of social practices is a simplified one, and they claim that they focus only on 'instances in which interpretations and symbolic associations are relatively uncontested', and play down the 'relative, situated and emergent' nature of the attribution of meaning (2012, p. 53). These caveats are not raised again as issues, so we must assume that they are not considered significant in developing the theory, or in the selected examples, including climate change and sustainability. However, anthropogenic ecological degradation and many other issues of importance and relevance to everyday life are surely 'instances' where 'interpretations and symbolic associations' are anything but uncontested. The reality of anthropogenic climate change is a 'relative, situated and emergent' meaning par excellence, reliant on the coherent collation, narration and reception of multiple scientific observations and models.

So too is that meaning highly contested, thanks to a cocktail that includes a surplus of cultural messages (Hamilton 2012); 'regime resistance' involving active opposition by powerful incumbent actors (Geels 2014); the manipulation of circulating discourses (Jacques et al. 2008); and a 'socially generated silence' in which individual and collective denial proliferates (Marshall 2014a). As individuals, groups, communities and societies we engage with this contestation of meaning, we contribute to it, process it, struggle with it in ways that we may or may not be fully conscious of. These processes have implications for personal and social life, and for subsequent responses to anthropogenic ecological degradation (Soron 2010). It could be argued that we carry on being recruits to 'high-resource' intensive practices regardless of this engagement, in which case, bypassing it may be pragmatic. However, the negotiation of meaning might be more central to the continuation or discontinuation of practices than a social practice approach concedes.

In fact there is a hollowing out of the dynamics involved in the production and negotiation of meaning in the application of a social practice approach in this context, where meaning becomes more or less synonymous with 'convention' (e.g. Hand et al. 2005, p. 6.7; Shove et al. 2012, p. 106, p. 132; Warde 2013); and listed as separate from 'expectation' but not differentiated from it (e.g. Shove et al. 2012, p. 153, p. 163). But these words have different if related definitions in everyday language and, normally, in the social sciences. A meaning is a denotation, referent, or idea that is linked to something—a word, term; or in the context of social practice, the 'symbolic significance' of a saying or doing. A convention is the way in which something is usually done; an expectation is the belief that something will happen, perhaps that something will be done as it is usually done in this context. A convention obviously relies on some form of 'agreement', but this might be unspoken, prefigured, tacit or spoken, conscious, deliberate; expected beforehand or deferred indefinitely (see Goffman 1971, p. 4).

Although not explicitly related to social practice theory, a recent analysis of social conventions is of interest here (Al-Amoudi and Latsis 2014, p. 15). In exploring how social conventions can be both arbitrary and normative, Al-Amoudi and Latsis present meaning and convention as in dynamic relation to each other, and in relation to competence, rather than as interchangeable terms. They describe how they intersect as follows:

> Every time a person adopts a convention, he or she also contributes to its reproduction; and conversely, by refraining from adopting a given convention, s/he contributes to its disappearance, replacement or, at least, transformation into a different convention. In effect, a single instance of transgressing a given convention can have shallower or deeper effects on the convention itself, depending on whether and how it is justified. When the breach is justified by reference to contingent reasons, the convention

itself is unlikely to be put into question. However, if the transgression is not justified by reference to accidental circumstances, and if the transgressor is deemed to be a competent participant acting legitimately, then the convention's normativity is challenged and is likely to become less and less respected as the number of accepted breaches increases. (2014, p. 15)

Up to the mention of a 'breach' their description is akin to social practice approach to how individuals are recruited into, or defect from social practices. Meaning is essential to a convention, which contributes as a base element to the reproduction of a social practice. But they then say that the *meaning* attributed to a breach is vital to whether a defection might have a ripple effect by encouraging others to defect. Recruitment to and defection from a social practice is dependent on 'how it is justified', in terms of contingent versus legitimate reasons, with associated interpretations of a transgressor's competence or otherwise, and the subsequent potential challenge to normativity.

Acknowledging the struggle over the meanings, competences and materials that contribute to a social practice opens a window onto 'a micro-politics of everyday life' (Sayer 2013, p. 169) that is largely missing from current conceptualizations of social practice in relation to sustainability (Schönian and Laube 2013, p. 126). As Sayer asserts 'an emphasis on norms, treated as no more than conventions, has the ironic effect of de-normativizing and de-politicizing conduct, by removing any notion of active judgement, evaluation or normative force; people just follow the norms' (2013, p. 171). Attempting to legitimize the transgression of a social convention is, in more straightforward language, to contest the meaning of what is considered normal. It indicates, therefore, a more complex link between agency and social practices (Groves et al. 2016).

Al-Amoudi and Latsis point to a more dynamic relationship between meaning and convention, and flesh out Sayer's critical assertion that 'conceptualizing individuals as 'carriers' of practices might represent them as passive, ignoring their dynamic, normative or evaluative relation to practices' (Sayer 2013, p. 170). 'Meaning' is something much more settled in applications of a social practice approach to sustainability; the corollary 'glue' of convention that combines with the requisite elements of materiality and competency, belittling agency 'by representing... understandings as stable and "internalized" by actors' (Sayer 2013, p. 170). Sayer perhaps goes too far in saying that a social practice perspective in this context readily 'leads to' behaviourism (2013, p. 170)—but it certainly raises questions about ethical and political implications of a social practice approach in which subjective engagement is marginalized. In bypassing the need to engage with people as mindful, deliberative subjects and relations, there is a clear danger that the 'service' of a social practice approach is to 'present policy makers with instrumentally-useful knowledge for manipulating practices so as to achieve external policy goals, whatever they may be (green or otherwise) over the heads of those involved in the practices' (Sayer 2013, p. 172). Ivan Illich, Andre Gorz, and many others, have long argued that the realization of environmentalism 'does not necessarily imply the rejection of authoritarian, technofascist solutions' (Gorz 1979, p. 17).[4]

Moving on

The most basic critical point to make here is that the relationship between agency and social practices remains complex, and as this brief discussion of how meaning is contested in relation to convention and competency indicates, one that is underexplored in recent developments of a social practice approach. The dynamics of subjective and intersubjective deliberations, explanations and judgments can be vital to 'recruitment' and 'defection', not least to 'sustainable' practices. For Groves et al. (2016), this means that other elements are involved that a practice theory of (un)sustainable behaviour struggles to address. They identify emotional attachments to, and investments in, practices intertwined with individual biographies, and ontological insecurities derived from risky and uncertain futures (2016, p. 4). Harrison similarly cautions against applying a social practice approach exhaustively to the diversity of human experience; when 'granting it such priority we can miss the fact that our lives take place as and through many states, only some of which can comfortably fit this grammar and category of action and doing' (Harrison 2009, p. 996). Of these 'many states' Harrison highlights embodied and affective experiences of loss, withdrawal and ambivalence which disrupt and unsettle any predictable flow of social practices; creating an 'interval or gap between doer and deed, practitioner and practice' (2009, p. 994) that social practice approaches struggle to articulate.

At the very least, it is vital to acknowledge that the 'sustainability' or otherwise of practices enters into the meaning of those practices for us; which we negotiate, defend, deny, embrace or struggle with some of the time (Soron 2010). Thus 'practices... matter to subjects... precisely because they help create expectations, produce feelings of autonomy by encouraging mastery of competences, and reinforce connection with others through shared meanings' (Groves et al. 2016, p. 8).[5] This recursive move is, presumably, what drives social practice theorists to be interested in sustainability as a key issue, to dedicate significant amounts of time and energy to the topic. Yet the assumption seems to be that outside of the academy, reflexive consideration of the reality of ecological crisis is unlikely to impact on what we say and do. Here we should be extremely wary: of exaggerating the power of reflexivity to transform the situations we find ourselves in; but also of avoiding arrogant assumptions about its exclusivity, whereby it is always 'others' who are more likely to be 'ventriloquized' by social norms or discourses (Sayer 2013, p. 172). Groves et al. rightly assert that considering neglected elements opens up 'a psycho-social route, distinct from practice theory though related to it, through which individual biographies may be linked to wider socio-cultural patterns of meaning and agency' (2016, pp. 3–4). In the next chapter this psychosocial route is explored in more detail; considering how we comprehend the reality of ecological crisis individually and collectively, and how these dynamics are related to the themes identified in this chapter.

Notes

1. See http://gofossilfree.org/uk/ and http://www.theguardian.com/environ-ment/series/keep-it-in-the-ground for more details.
2. Klein's account of the powerful international social network – 'a geoclique' – linking venture capitalists, climate scientists and engineers to promote 'geoengi-neering' solutions to climate change; and of the links between lobbyists, government departments, and the oil and gas industry in the UK, are specific examples of alliances directly relevant to anthropogenic ecological degradation (Klein 2014). See in particular Chapter 4 'Planning and banning' (pp. 120–160) and Chapter 8 'Dimming the sun' (pp. 256–292). In terms of the UK government and energy company alliances, Klein notes that 'in the UK, the energy industry met with Department of Energy and Climate Change roughly eleven times more frequently than green groups did during David Cameron's first year

in office. In fact it has become increasingly difficult to discern where the oil and gas industry ends and the British government begins. As *The Guardian* reported in 2011, 'At least 50 employees of companies including EDF Energy, npower and Centrica have been placed within government to work on energy issues in the past four years... The staff are provided free of charge and work within the departments for secondments of up to two years' (Klein 2014, p. 149).

3. More worrying perhaps is Klein's point that rather than demanding consensual legitimacy, failing to transition to a genuinely sustainable society simply requires that we do nothing (Klein 2014).

4. The desirable alternative, for both Gorz and Illich, were the tools of 'conviviality' (Illich 1973): localized organization generating individual and collective autonomy. This kind of ecology opposes totalitarianism of left and right, and 'embodies the revolt of civil society and the movement for its reconstruction' (Gorz 1979, p. 40). Naomi Klein's recent contribution to the sustainability agenda is perhaps more closely allied with this earlier approach (Klein 2014), alongside broader creative responses such as the Dark Mountain Project. Of course, how such localized responses might 'break out' from beyond their current footholds remains a pertinent question.

5. Ironically, considering social practice theory's reliance on his theorization of social change, Giddens's later work considered the moral and existential questions associated with ecological crisis to be a good example of the 'return of the repressed' in contemporary 'post-traditional societies' (Giddens 1990, 1991, 1994). The establishment of modernity meant embedding dominion over the rest of nature as taken-for-granted, and incorporated it into material and social progress. 'Post-traditional society' strips such certainties away, and the ecological crisis, as in many areas of culture, is experienced as a profound ontological security, inhabiting practices with novel recursiveness, undermining practical and embodied orientations. As is well known, for Giddens this extended reflexivity reaches into the constitution of both self and society (Giddens 1991).

References

Al-Amoudi, I., & Latsis, J. (2014). The arbitrariness and normativity of social conventions. *The British Journal of Sociology, 65*(2), 358–378.

Avelino, F., & Rotmans, J. (2009). Power in transition: An interdisciplinary framework to study power in relation to structural change. *European Journal of Social Theory, 12*(4), 543–569.

Berners-Lee, M., & Clark, D. (2013). *The burning question. We can't burn half the world's oil, coal and gas. So how do we quit?* London: Profile Books.

Capstick, S., Whitmarsh, L., Poortinga, W., Pidgeon, N., & Upham, P. (2015). International trends in public perceptions of climate change over the past quarter century. *Wiley Interdisciplinary Reviews: Climate Change, 6*(1), 35–61.

Crist, E. (2004). Against the social construction of nature and wilderness. *Environmental Ethics, 26,* 5–24.

Cudworth, E. (2015). Killing animals: Sociology, species relations and institutionalized violence. *The Sociological Review, 63*(1), 1–18.

Dunlap, R. E., & McCright, A. M. (2010). Climate change denial: Sources, actors and strategies. In C. Lever-Tracy (Ed.), *Routledge handbook of climate change and society* (pp. 240–259). London: Routledge.

Dunlap, R. E., & McCright, A. M. (2011). Organized climate change denial. In J. S. Dryzek, R. B. Norgaard, & D. Schlosberg (Eds.), *The Oxford handbook of climate change* (pp. 144–160). Oxford: Oxford University Press.Fisher, A. (2013a, 2nd ed.) *Radical ecopsychology.* London: Routledge.

Fisher, A. (2013a), 2nd ed. Radical ecopsychology. London: Routledge.

Fisher, A. (2013b). Ecopsychology at the crossroads: Contesting the nature of a field. *Ecopsychology, 5*(3), 167–176.

Franklin, A. (1999). *Animals and modern culture: A sociology of human-animal relations in modernity.* Thousand Oaks: Sage.

Frantz, C. M., & Mayer, F. S. (2014). The importance of connection to nature in assessing environmental education programs. *Studies in Educational Evaluation, 41,* 85–89.

Geels, F. W. (2002). Technological transitions as evolutionary reconfiguration processes: A multi-level perspective and a case-study. *Research Policy, 31*(8–9), 1257–1274.

Geels, F. W. (2012). A socio-technical analysis of low-carbon transitions: Introducing the multi-level perspective into transport studies. *Journal of Transport Geography, 24,* 471–482.

Geels, F. W. (2014). Regime resistance against low-carbon transitions: Introducing politics and power into the multi-level perspective. *Theory, Culture and Society, 31*(5), 21–40.

Geels, F. W., & Schot, J. W. (2007). Typology of sociotechnical transition pathways. *Research Policy, 36*(3), 399–417.

Giddens, A. (1990). *The consequences of modernity.* Cambridge: Polity Press.

Giddens, A. (1991). *Modernity and self-identity.* Cambridge: Polity Press.

Giddens, A. (1994). Living in a post-traditional society. In U. Beck, A. Giddens, & S. Lash (Eds.), *Reflexive modernization* (pp. 56–109). Cambridge: Polity Press.

Goffman, E. (1971). *Relations in public: Microstudies of the social order.* London: Allen Lane.

Gorz, A. (1979). *Ecology as politics.* London: Southend Press.

Grin, J. (2010). Understanding transitions from a governance perspective. In J. Grin, J. Rotmans, J. Schot, F. W. Geels, & D. Loorbach (Eds.), *Transitions to sustainable development: New directions in the study of long term transformative change* (pp. 249–319). London: Routledge.

Groves, C., Henwood, K., Shirani, F., Butler, C., Parkhill, K., & Pidgeon, N. (2016a). Invested in unsustainability? On the psychosocial patterning of engagement in practices. *Environmental Values, 25*(3), 309–328.

Hamilton, C. (2012b). What history can teach us about climate change denial. In S. Weintrobe (Ed.), *Engaging with climate change: Psychoanalytic & interdisciplinary prespectives* (pp. 16–32). London: Routledge.

Hand, M., Shove, E., & Southerton, D. (2005). Explaining showering: A discussion of the material, conventional, and temporal dimensions of practice. *Sociological Research Online, 10*(2). http://www.socresonline.org.uk/10/2/hand.html. Accessed 18 Dec 2015.

Harrison, P. (2009). In the absence of practice. *Environment & Planning D: Society And Space, 27*(6), 987–1009.

Hoffman, A. J. (2011). Talking past each other? Cultural framing of skeptical and convinced logics in the climate change debate. *Organization and Environment, 24*, 3–33.

Hollway, W. (2016, January 19). Cited from personal correspondence with author.

Jacques, P., Dunlap, R. E., & Freeman, M. (2008). The organization of denial: Conservative think tanks and environmental scepticism. *Environmental Politics, 17*, 349–385.

Keister, L. A. (2014). The one percent. *Annual Review of Sociology, 40*, 347–367.

Kern, F. (2011). Ideas, institutions and interests: Explaining policy divergence in fostering 'system innovations' towards sustainability. *Environment and Planning C: Government and Policy, 29*(6), 1116–1134.

Khan, S. R. (2011). *Privilege: The making of an adolescent elite at St. Paul's school.* Princeton: Princeton University Press.

Khan, S. R. (2012a). Elite identities. *Identities, 19*(4), 477–484.

Khan, S. R. (2012b). The sociology of elites. *Annual Review of Sociology, 38*, 361–377.

Khan, S., & Jerolmack, C. (2013). Saying meritocracy and doing privilege. *The Sociological Quarterly, 54*(1), 9–19.

Khoo, S. (2013). Sustainable development of what? In F. Fahy & H. Rau (Eds.), *Methods of sustainability research in the social sciences.* London: Sage.

Klein, N. (2014). *This changes everything: Capital vs the climate.* Harmondsworth: Penguin.

Lakoff, G. (2010). Why it matters how we frame the environment. *Environmental Communication, 4*(1), 70–81.

Leary, M. R., & Schreindorfer, L. S. (1997). Unresolved issues with terror management theory. *Psychological Inquiry, 8*(1), 26–29.

Levy, D. L., & Newell, P. (2002). Business strategy and international environmental governance: Toward a neo Gramscian synthesis. *Global Environmental Politics, 2*(4), 84–101.

Maniates, M. (2014). Sustainable consumption – Three paradoxes. *GAIA, 23/S1*(2014), 201–208.

Marshall, G. (2014a). *Don't even think about it: Why our brains are wired to ignore climate change.* London: Bloomsbury.

Marshall, G. (2014b). Five. In J. Smith, R. Tyszczuk, & R. Butler (Ed.), *Culture and climate change: Narratives* (Vol. 2, pp. 96–97). Cambridge: Shed.

Randall, R. (2009). Loss and climate change: The cost of parallel narratives. *Ecopsychology, 3*, 118–129.

Rustin, M. (2013). How is climate change an issue for psychoanalysis? In S. Weintrobe (Ed.), *Engaging with climate change: Psychoanalytic and interdisciplinary perpsectives* (pp. 170–185). London: Routledge.

Sayer, A. (2013). Power, sustainability and well-being: An outsider's view. In E. Shove & N. Spurling (Eds.), *Sustainable practices: Social theory and climate change* (pp. 292–317). London: Routledge.

Sayer, A. (2014). *Why we can't afford the rich.* Bristol: Policy Press.

Schönian, K., & Laube, S. (2013). Book review: The dynamics of social practice: Everyday life and how it changes. *Science and Technology Studies, 26*(3), 124–126.

Shove, E. A., Pantzar, M., & Watson, M. (2012). *The dynamics of social practice: Everyday life and how it changes.* London: Sage.

Shove, E., & Walker, G. (2014). What is energy for? Social practice and energy demand. *Theory, Culture & Society, 31*(5), 41–58.

Smith, A., Stirling, A., & Berkhout, F. (2005). The governance of sustainable sociotechnical transitions. *Research Policy, 34*(10), 1491–1510.

Solomon, S., Greenberg, J., & Pyszczynski, T. (1991). A terror management theory of social behavior: The psychological functions of self-esteem and cultural worldviews. *Advances in experimental social psychology, 24*, 93–159.

Soron, D. (2010). Sustainability, self-identity and the sociology of consumption. *Sustainable Development, 18*(3), 172–181.

Turnheim, B., & Geels, F. W. (2013). The destabilisation of existing regimes: Confronting a multi-dimensional framework with a case study of the British coal industry (1913–1967). *Research Policy, 42*(10), 1749–1767.

Vining, J. (2003). The connection to other animals and caring for nature. *Human Ecology Review, 10*(2), 87–99.

Warde, A. (2013). Sustainable consumption and behaviour change. *Discover Society, 1*.http://www.discoversociety.org/2013/10/01/sustainable-consumption-and-behaviour-change/. Accessed 18 Dec 2015.

Wilk, R. (2010). Consumption embedded in culture and language: Implications for finding sustainability. *Sustainability: Science, Practice, & Policy, 6*(2), 38–48.

6

Managing Terror: Mortality Salience, Ontological Insecurity and Ecocide

Introduction

> [U]ntil the late twentieth century, every generation throughout history
> lived with the tacit certainty that there would be generations to follow.
> Each assumed, without questioning, that its children and children's chil-
> dren would walk the same Earth, under the same sky... that certainty is
> now lost to us... that loss, unmeasured and immeasurable, is the pivotal
> psychological reality of our time. (Macy, cited in Mnguni 2010, p. 125)

In the previous chapter, we addressed some of the potential limitations of
a social practice perspective, in paying relatively little attention to issues
of meaning, emotion, power and the relationship between humans and
the rest of nature. Alternative perspectives grant a much more central
role to these dynamics, in making sense of human engagement with the
anthropogenic ecological crisis. In pursuing this book's themes then,
there is clearly value in considering what these alternatives might contribute to our understanding.

© The Author(s) 2016
M. Adams, *Ecological Crisis, Sustainability and the Psychosocial Subject*,
DOI 10.1057/978-1-137-35160-9_6

This and the following chapter are devoted to a number of perspectives that explicitly address the complex relationship between 'knowing' and 'not-knowing' about the anthropogenic ecological crisis. Unlike a social practice theory, these approaches emphasize the importance of embodiment and emotion, and the subjective and intersubjective negotiation and contestation of meaning. Although varied, the perspectives we will consider share a single assumption —that awareness of anthropogenic ecological crisis generates profound and unconscious anxiety. Hence, our initial task here is to say a little about why this is thought to be the case. It may be obvious for the most part, but it is also a way of introducing a number of concepts that will run through the rest of the chapter, particularly, ontological security and mortality salience.

Ontological (In)security

Borrowing heavily from Laing (1960), Erikson (1950) and others, Giddens' sociological perspective on ontological (in)security is still influential today (Giddens 1990, 1991, 1992).[1] For Giddens, ontological security is derived from a more generalized 'basic trust' in the continuity of everyday life, and in a sense of self and others that maintains that continuity. It refers to a 'person's fundamental sense of safety in the world and includes a basic trust of other people. Obtaining such trust becomes necessary in order for a person to maintain a sense of psychological well-being and avoid existential anxiety' (Giddens 1991, pp. 38–39). This trust is understood to be first formed in the consistent care and reciprocity of relationships with early caregivers. Basic trust involves a faith in the consistency and permanence of others initiated in childhood but is also maintained by interpersonal, communal and cultural moorings. Giddens makes an important conceptual move here, mutually implicating psychic, interpersonal and sociocultural dynamics in a way that chimed with a number of his contemporaries (e.g. Bruner 1986; Gergen 1991).

Trust allows the 'bracketing off' of questions about the meaning and purpose of existence and one's place in the world by providing a sense of shared normality. This socially constructed and conciliated 'common sense' provides a way of 'answering' existential questions that become tacit and taken-for-granted in our everyday orientation to the world: 'The ontologically secure individual does not worry about the meaning of life, or of his or her life, or of its purpose; s/he does not worry about the social world collapsing' (Croft 2012, p. 221). In this way, ontological security involves the provision of meaning, both as an unconscious embodied experience anchored in infancy and as non-conscious experience or 'practical consciousness' of taken-for-granted routines and practices (Giddens 1991).

Ontological *in*security, by contrast, is described as a state where one 'cannot take the realness, aliveness, autonomy and identity of [one]self and others for granted' (Laing 1960, p. 141). This uncertainty is normally described as profoundly unsettling—inducing anxiety and dread (Giddens 1990, p. 97; Croft 2012, p. 222). Ontological insecurity is, therefore, 'the fear of being overwhelmed by anxieties that reach to the very roots of our coherent sense of 'being in the world'' (Giddens 1991, p. 37).

Although the shared framework of reality upon which ontological security relies is culturally, historically and geographically contingent, the need for ontological security is considered to be universal and primary: 'Humans, thus, are assumed to have a basic need to understand the world in a way that protects them from doubts and allows them to ascribe meaning to their existence' (Van Marle and Maruna 2010, p. 9). It is within these shared cultural frameworks of reality that we construct our own personal narratives, and imbue them with desire, meaning and value. Thus, both ontological security and ontological insecurity is understood as a fundamentally psychosocial process, embedded in collective forms of relating and meaning-making, while affectively and cognitively embodied.

Clinical studies indicate how individual trauma can threaten ontological security, such as when we experience the death of significant others (Mellor and Shilling 1993); pulling at the fabric that allows us to routinely 'go on' (e.g. Yip 2004). Collective traumas can also tear at the shared, culturally derived, but often unspoken, ways of 'going on' that provide that security (Alexander 2004; Zaretsky 2002). Claims about the elements of a culture that provide grounds for ontological security are varied, but a number of social theorists have argued that the transition to 'late modernity' is akin to an ongoing collective trauma, ushering in an era of psychosocial fragmentation in which community, collective and public sphere is threatened alongside established traditions, conventions and narratives that help us to navigate life and death meaningfully (Bracken 2001).[2] In fact, the apparently fraught search for meaning and security and associated complex of anxieties accompanying late modernity is often referred to collectively *as* ontological insecurity (Van Marle and Maruna 2010, p. 8).

Ecological Crisis and Ontological Insecurity

Sociological concepts of ontological security tell us that a taken-for-granted sense of imperturbability is an ongoing accomplishment and relatively fragile. In this sense, the communication of anthropogenic ecological crisis threatens the imperturbability of everyday psychological reality, as Joanna Macy's words at the beginning of this chapter suggest. At the same time, anthropogenic ecological crisis undermines a number of related 'certainties' that have come to provide a taken-for-granted foundation for day-to-day existence for many: 'Narratives of anthropogenic climate change threaten implicit trust in some of the foundations of late modern society, including consumerism, individual freedom, capitalism, and liberal democracy' (Lucas et al. 2015, p. 85). According to the framework articulated here, an ongoing *psychosocial* organization is required to maintain a semblance of ontological security, an issue we return to below.

Establishing an empirical base for an assertion of culture-wide ontological insecurity is difficult, perhaps impossible. Ascertaining its roots in awareness (at some level), or a felt sense of ecological crisis, is even more prone to conjecture and uncertainty, especially if effective responses are considered to be unconscious or only partially conscious. It has been notoriously difficult to build empirically from Giddens' own analysis, for example, despite his impressive meta-theoretical foundations (Pozzebon and Pinsonneault 2005). Nonetheless, related factors such as the rise in anxiety disorders and depression, loneliness and isolation, and self-harm in relatively wealthy nations (e.g. Coughlan et al. 2014; Griffin 2010; Layard 2006; Ness et al. 2015), is taken to suggest malaise and melancholy at a cultural level, often associated with the limits of materialism to provide the basis for a meaningful shared existence (Verhaeghe 2014).

To state it bluntly, we are always likely to be on uncertain ground in forwarding an argument about how lots of people feel, even if they deny it; or that people's actions are shaped by the unfathomable reality of their own mortality, even if they are not aware of it. However, there is a rich history in existential philosophy and psychoanalysis that grapples with these issues. While it is undoubtedly true that they sit uneasily with the 'factors and variables' approach of mainstream psychology and social psychology, the logic of existential and applied psychoanalytic perspectives can have a deep intuitive appeal, as exemplars of few theoretical approaches embracing the uncanny, irrational and complex nature of human socioemotional relations (Craib 1998). One such perspective, Terror Management Theory, attempts to develop an existential approach that also draws on psychoanalytic understandings, while being at least partially located in the empirical tradition of experimental social psychology.

Terror Management Theory and Mortality Salience

Advocates of Terror Management Theory offer a parallel existential perspective on the impact of ecological crisis and potentially provide further insight into the threat it poses to ontological security.[3] Terror Management Theory is based on the account of the relationship between death and the human psyche proposed by the anthropologist Ernest Becker (Becker 1973). For Becker, fear of annihilation, finitude and nothingness are considered a universal and primary source of anxiety, or, to use our preferred term, ontological insecurity.

Bauman follows Bakhtin in describing this experience as 'cosmic fear': 'a *human*, all-too human emotion aroused by the unearthly, *inhuman* magnificence of the universe' (Bauman 2004, p. 71; emphasis in the original). Cosmic fear encompasses a sense of fright, vulnerability, insignificance, but also a brings us face-to-face with a fundamentally unsettling ambiguity: 'the realization that it is not in the power of humans to grasp, comprehend, mentally assimilate that awesome might which manifests itself in the sheer grandiosity of the universe' (ibid). Bauman cites Pascal's 'flawless' description of that feeling:

> When I consider the brief span of my life absorbed into the eternity which comes before and after... the small space I occupy and which I see swallowed up in the infinite immensity of spaces of which I know nothing and which knows nothing of me, I take fright and am amazed to see myself here rather than there, now rather than then. (Cited in Bauman 2004, pp. 71–2)

While many others have highlighted the significance of death to the construction of personal and social life (e.g. Berger 1969; Frankl 1963; Lifton 1979), Becker's specific assertion is that human beings' knowledge of their own mortality is so profoundly unsettling that it amounts to a universal psychological predisposition 'to suppress thoughts of death to manage anxiety about the inevitability of mortality…. thinking about death is so costly that denial of death is ubiquitous and explains the majority of human mythologies and world views' (Dickinson 2009, p. 35).

The mythologies and worldviews highlighted by Becker are the 'anxiety-buffering' vehicles through which we seek to establish a sense of meaning, permanence and self-esteem (Pyszczynski et al. 2006). They are akin to the role attributed to shared narratives and frameworks of reality in maintaining ontological security in sociological accounts: the design of 'elaborate subterfuges' by human societies (Bauman 1992, p. 1). Making sense of the reality of death is perhaps the ultimate 'existential question' we seek to bracket off via socially and culturally validated practices. Terror Management Theory adds the ingredient of 'dread' as a motivational predisposition for seeking ongoing ontological security; and a significant amount of behaviour as a consequence. In sum, 'the potential for abject terror created by the awareness of the inevitability of death in an animal instinctively programmed for self-preservation and continued experience lies at the root of a great deal of human motivation and behavior' (Greenberg et al. 1997, p. 61).

Such claims do not, in themselves, give any clue as to how we respond to this apparently deep disquiet. According to an existential framework, it can be positive or negative in terms of human flourishing and meaning (Vail et al. 2012). In fact, from an existentialist perspective more broadly, awareness of death is a necessary prerequisite to the discovery of meaning (Pienaar 2011). For Terror Management Theory, the predominant human response to knowing about death is its 'denial' through the shared pursuit of 'symbolic selves' that, unlike physical selves, can transcend mortality. 'Fictional selves' are co-constructed in cultural and symbolic systems—shared worldviews that outlive and symbolically extend our finite embodied selves (Greenberg et al. 1997). Investing meaning in an idealized 'other' can also reduce death anxiety. In transferring our hopes and fears to idealized others, we also attempt to transcend our own finiteness and bolster a symbolic self. The projection of power onto leaders, deities or celebrities endows the venerated other with the power to absorb or 'save' us. The problem of heroics' is not just a facet of celebrity worship or similar cultural froth we can readily disparage in others; for Becker, it is 'is the central one of human life... Society is itself a codified hero system, which means that the society everywhere is a living myth of the significance of human life, a defiant creation of meaning' (Becker 1973, p. 7). Human culture is, in effect, an 'immortality-striving hero system' (Dickinson 2009, p. 40); a striving for 'symbolic immortality' in the face of physical finiteness (Lifton 1979); or what Seale refers to suggestively as 'resurrective practices' (Seale, 2002).[4] Paradoxically, Becker pointed out that 'immortality-striving' behaviour triggered by mortality salience might actually risk harming one's physical self. This is possible if behaviour, despite harming or risking harm to oneself (e.g. smoking), contributes to the investment in a symbolic self that contributes to an 'immortality-striving hero system' (e.g. rebel, sophisticate).

Terror Management Theory reflects a concerted effort to fit Becker's ideas into the mould of experimental social psychology. To this end, it has established sizeable empirical support and undergone theoretical development (see Greenberg et al. 1997; Burke et al. 2010 and Hayes et al. 2010 for overviews); although it is not uncontested (e.g. Kirkpatrick and Navarrete 2006; Navarrete and Fessler 2005). Experiments commonly involve exposing participants to verbal or written 'death-primes', which are designed to manipulate 'mortality-salience' or a more general 'death thought accessibility' (Hayes et al. 2010)—imagining one's own death, images involving deaths of others, or exposure to objects and places associated with death such as funeral homes (McCabe et al. 2016). Findings suggest that participants who are 'death-primed' can be more likely to engage in stereotypic thinking (Schimel et al. 1999), and more positively value driving recklessly (Ben-Ari et al. 1999); smoking (Hansen et al. 2010; Martin and Kamins 2010), and binge drinking (Jessop and Wade 2008)—if the behaviour is already tied in with their self-esteem and adopted worldview. Accordingly, in applying Terror Management Theory to health sciences, a common research finding suggests that where health promotion campaign text or imagery makes mortality salient (e.g. related depictions of death or dying), they may actually 'precipitate the very behaviours which they aim to deter among some recipients' (Jessop and Wade 2008, p. 773). Here we see Terror Management Theory's practical illustration of Becker's suggestion that we may seek behaviour that risks death or harm to our physical self, if it maintains our 'symbolic self' in a codified hero system.

The communication of anthropogenic ecological crisis is considered significant to Terror Management Theory because it potentially increases 'death thought accessibility'; it reminds us, simply put, of our own mortality, and can, therefore, generate the varied responses to 'death-primes' that Terror Management Theory has identified (Gifford 2011). The most comprehensive discussion of the potential of Terror Management Theory in this regard is Dickinson's (2009). Her initial prediction is that in communicating future impacts of climate change, the greater the severity of the forecast, the greater the accessibility of thoughts of death:

> I know of no study that investigated whether delivering information on global climate change increases death thought accessibility. However, if it does, then experiments that manipulate the way in which information on global climate change is presented, including the extent to which graphic details or the potential for human mortality are revealed, could prove useful not only for testing the idea that mortality salience influences human response to global climate change but also for determining the most effective ways to structure climate change education. (Dickinson 2009, p. 5)

At the time of writing, as far as I am aware, there are still no published studies of climate change communication as a trigger for mortality salience; nor are climate change and ecological crisis appearing in calls for Terror Management Theory to be applied to a range of social issues (e.g. Arndt and Vess 2008). Broadly related research to date has been occupied with the extent to which manipulations of mortality salience affect pro-environmental attitudes and intentions; materialistic values and conspicuous consumption (Arndt et al. 2004); and how contemplation of death impacts on people's understanding of the relationship between humans and the rest of nature (e.g. Fritsche et al. 2010; Fritsche and Häfner 2012; Goldenberg et al. 2001; Kasser and Sheldon 2000; Koole and Van den Berg 2005; Vess and Arndt 2008). Here we can only speculate further as to why anthropogenic ecological crisis might operate as a 'death prime', assisted by reference to the concept of ontological security as introduced earlier.

Extending Death Accessibility

Anthropogenic ecological crisis makes accessible thoughts about death or serious harm in relation to self and others, including future generations. We might add that 'in an era when humanity has the capacity to destroy all life' (Hollander 2009, p. 3) 'death accessibility' is extended—if we contemplate death or harm to other species, natural habitats or places we have formed attachments to, or the planet in a more general sense as a 'home'. Kinnvall's discussion of the struggle to maintain a sense of identity in an era of globalization, diaspora, refugees, and migration provides a helpful illustration of this point (2004, p. 747). She uses Giddens' understanding to make sense of migrant experiences of ontological insecurity in relation to a sense of home:

> The very category of "home" as a bearer of security can be found in its ability to link together a material environment with a deeply emotional set of meanings relating to permanence and continuity.... Ontological security is maintained when home is able to provide a site of constancy in the social and material environment. Home, in this sense, constitutes a spatial context in which daily routines of human existence are performed. It is a domain where people feel most in control of their lives because they feel

free from the social pressure that is part of the contemporary world. Home, in other words, is a secure base on which identities are constructed.... Homelessness is exactly the opposite, as it is characterized by impermanence and discontinuity. (Kinnvall 2004, p. 747)

Anthropogenic ecological crisis is already a significant driver of migration and the movement of people (Biermann and Boas 2008), although it is resisted (McNamara and Gibson 2009), so there is a straightforward connection to Kinnvall's point in that sense. There is also a less obvious connection. Kinnvall's argument is based on the loss of security invested in 'home' resulting from the rapid socioeconomic changes associated with globalization, and the movement of people that follows. But there is also a sense in which anthropogenic ecological crisis unsettles the category of home more widely, questioning its permanence and continuity, be it in response to new methods of resource extraction, current degradation, or as a mediated future threat. Albrecht refers evocatively to these ambivalent place attachments as 'solastalgia' (Albrecht 2012)—'the homesickness you have when you are still at home'. The communication of ecological crisis, particularly when it is linked to severe material, social and personal impacts, unsettles the implicit faith in the world characteristic of ontological security: 'Climate change research demands that we re-evaluate our trust in many elements of our everyday lives in a way that is profoundly unsettling' (Lucas et al. 2015, p. 79).

Terror is elicited here not via literal 'death' thoughts, but an existential 'dread' associated with the stark reminder of 'human creatureliness', finitude and insignificance (Cox et al. 2007; Last 2013)—unravelling the 'immortality-striving' premise of the 'symbolic self' described earlier.[5] As such striving is anchored to contingent interpersonal, social and cultural moorings it closely parallels arguments made about the development and maintenance of ontological security.[6] Ontological security normally offers 'a protection against future threat and dangers which allows the individual to sustain hope and courage in the face of whatever debilitating circumstances she or he might later confront' (Giddens 1991, p. 39); part and parcel of managing that 'threat' is 'the ongoing effort to cope with the anxiety associated with death and annihilation' (Vandenberg 1991, p. 1279).

Beyond Terror

There are undoubtedly reductive implications in Terror Management Theory's focus on the terror of death as *the* 'master motive' of human behaviour, as a number of critics have pointed out (Lerner 1997; Muraven and Baumeister 1997; Pelham 1997; Vallacher 1997). One line of criticism points to the obvious possibility that people have other primary concerns that can trigger anxiety at similar or greater levels than a primeval dread of biological death—what Solomon et al. refer to as 'absolute annihilation' (1991, p. 96). These might include uncertainty or the unknown (McGregor et al. 2001), the threat of meaninglessness (Heine et al. 2006); separation from loved ones; or even eternal damnation (Leary and Schreindorfer 1997, p. 28) —rather than a fear of death per se. Another is to question neglect of the role of more conscious thoughts and concerns about death (Arndt and Vess 2008); and, we might add, the ongoing impact of narrative and discursive constructions.

A Terror Management framework can accommodate cultural variation in the ways in which we respond to mortality being made salient. If there is a tendency to defend one's word view and self-esteem, it follows that, as cultural worldviews vary, so will defences—in fact culturally divergent responses are to be expected. A culture defined by the acquisition of goods will lead to defences different to one in which asceticism is valued highly. Literature addressing cross-cultural comparisons are relatively sparse, but do suggest some support for the claim that while defences vary, what they share is a defence of cultural worldview (Fernandez et al. 2010; Ma-Kellams and Blascovich 2011; Routledge et al. 2010; Wakimoto 2006).[7]

However, other research questions any straightforward links between social and cultural norms and mortality salience when studied in cross-cultural context (Du and Jonas 2015; Yen 2013). This leads to a potentially more unsettling critique, not about the specificity of responses to fear of death, but about whether the fear of death itself is culturally specific. Such a possibility is hinted at in a meta-analysis of terror management studies (Burke et al. 2010). The authors found that while the effects of mortality salience manipulations on world view and self-esteem defence are moderately significant across almost 300 experiments, they affected Americans significantly more than Europeans or Asians (Burke et al. 2010, p. 182).

The authors consider a possible explanation—that the 'idea of death' is relatively less integrated into American culture, so when Americans are prompted to contemplate death, they are more defensive. More broadly, they suggest that regional differences indicate how 'cultural factors may significantly alter how people's insecurities about death manifest themselves, indicating the methodological importance of researchers taking cultural climate into account when constructing their dependent measures' (Burke et al. 2010, p. 182).[8] There is a more profound theoretical implication here though, that the authors do not consider: a problematization of fear of death as universal master motive. World views, especially where they intersect with tradition, religion and spirituality, are often explicitly concerned with death and meaning (Berger 1973; Park 2005).

We need not accept the motivational primacy of a terror of death and attempts to manage it, as *the* existential theme significant in shaping human behaviour and experience, to take something of value from Terror Management Theory here. Whatever its shortcomings, the theory brings to the forefront the importance of how we 'answer' existential questions in everyday life (Giddens 1991), and, vitally, their intersection with both embodied affective and cognitive dynamics and sociocultural frameworks for navigating through life. We can move tentatively from the enticing but reductive focus on the 'master motive' of death awareness, to the broader concept of ontological (in)security. In this light, the implications of Terror Management Theory for how we search for and construct *meaning* is central to what matters about how we cope with terror of death, but also, perhaps more constructively, how we embrace life-giving—our generative powers (including reproduction and birth)—and relational attachments (Cox et al. 2008; Fritsche et al. 2007; Mikulincer et al. 2003). As Wendy Hollway points out, in asserting that contingent cultural frameworks are needed to transcend the fear of death, Terror Management Theory reveals a 'masculinist' bias: 'it forgets that women, through the creation of life and the future of the human species, have an ineffable way of transcending death' (Hollway 2016). Generative and relational capabilities might potentially inform master narratives that meaningfully address awareness of the inevitability of death *and* embrace the perpetuation of life.

At least two related premises are worth pursuing further in the context of understanding responses to anthropogenic ecological degradation. First, that when threats and dangers penetrate the emotional inoculation of ontological security, existential anxiety surfaces; second, that anxiety is commonly dealt with through shared social and cultural frameworks that are the basis for the shared meaning-making that symbolically transcends death. Alongside Terror Management Theory, in fields such as sociology, social psychology and social or applied psychoanalysis, a disparate body of work has emerged in recent years researching people's responses to knowledge of anthropogenic ecological crisis, studying what people say as evidence of underlying anxiety (e.g. Hanson-Easey et al. 2015; Olausson 2011; Jaspal et al. 2012; Threadgold 2012; Stoll-Kleeman et al. 2001). This work is addressed in the next chapter.

Notes

1. The critical psychiatrist R.D. Laing, initially utilized the concept of ontological *in*security as part of his understanding of the processes underlying schizoid and psychotic states (1960). Laing conceptualizes it as a sense of self that is chronically dependent on the contingent affirmation of others, and therefore, precarious, detached and hollow. He argued that experiences in infancy or early childhood were central to the development of a more or less secure sense of self. Laing follows the tradition of Winnicott and others, in emphasizing the importance of the relational, 'transitional space' between a child and their primary carers (Winnicott 1965, 1974). If the latter are responsive enough to the child's needs and feelings, she will likely develop a firm sense of what those needs are, come to recognize them as her own, and in parallel establish a fledgling distinction between self and not-self—the basis for ontological security. On the other hand harsh, inconsistent care, or the persistent casting of the child's needs to fit with the expectations and demands of the caregiver, encourages the development of uncertain boundaries between self and other—ontological insecurity. This all happens at an unconscious level and is argued to impact on the nascent psyche as a profoundly formative experience.
2. Some sociologists, including Giddens, are cautiously positive about the degree of autonomy and freedom this uncertainty and disembeddedness generates; others, such as Bauman, are less convinced (e.g. Bauman 1991). See Adams (2007) for an extended discussion of various takes on what is referred to as the 'psychosocial fragmentation thesis'.
3. See Van Marle and Maruna (2010) for a more detailed account of the linkages between these two concepts.
4. For Seal 'the body and its death' is the root from which most forms of social life arise. In a similar, though simpler, sense to Becker he argues that we con-

struct narratives that orient us 'towards life and away from death' and these are collectively referred to as 'resurrective practices'. Informal everyday resurrective practices 'have in common an affirmation of the social bond in the face of its dissolution, enabling people to claim membership in an imagined human community' (Seale 2000, p. 36). Though broadly parallel, Becker's account is more ambivalent about the affirmative nature of social and cultural life as response to the reality of death.

5. According to the logic of Terror Management Theory, fear of 'human creatureliness' is an extension of the fear of death:

> If humans manage the terror associated with death by clinging to a symbolic cultural view of reality, then reminders of one's corporeal animal nature would threaten the efficacy of this anxiety-buffering mechanism. As argued by Becker, the body and its functions are therefore a particular problem for humans. How can people rest assured that they exist on a more meaningful and higher (and hence longer lasting) plane than mere animals, when the sweat, bleed, defecate, and procreate, just like other animals? (Goldenberg et al. 2002, p. 2)

This hypothesis has been borne out to some extent in Terror Management studies where partipants are reminded of their own physicality in various ways, and responding with disdain or disgust when mortality is made salient (Cox et al. 2007; Goldenberg 2005). This hypothesis is also connected with wider expressions of misogyny and the denigration of women as somehow 'closer' to nature (Goldenberg and Roberts 2004). It has fascinating connotations for our later discussion of what symbolism and mythology a genuinely 'sustainable' cultural worldview might need to embrace. Perhaps it involves turning the derogation of human creatureliness on its head –celebrating and valorizing creatureliness and our nonhuman connections.

6. The experience is also similar to that described by Berlant, with a psychoanalytic emphasis, as contributing to 'cruel optimism': 'if a relation in which you've invested fantasies of your own coherence and potential breaks down, the world itself feels endangered' (Berlant 2012).

7. Kashima et al. (2004) studied mortality salience in relation to individualism, comparing cultures which are, broadly speaking, individualist in world view (Australia) or collectivist (Japan). Their results supported their hypothesis that mortality salience enhanced a defence of individualism in Australian participants, but reduced it in Japanese participants.

8. Sociologists have long argued that late modern societies, of which we might consider American society the epitome, increasingly sequester experiences of death from everyday life (e.g. Giddens 1991; Mellor and Shilling 1993).

References

Adams, M. (2007). *Self and social change*. London: Sage.

Albrecht, G. (2012). The age of solastalgia. *The Conversation, 7*. https://theconversation.com/the-age-of-solastalgia-8337. Accessed 18 Dec 2015.

Alexander, J. C. (2004). Toward a theory of cultural trauma. In R. Eyerman, J. C. Alexander, B. Giesen, N. J. Smelser, & P. Sztompka (Eds.), *Cultural trauma and collective identity* (pp. 1–30). California: University of California Press.

Arndt, J., & Vess, M. (2008). Tales from existential oceans: Terror management theory and how the awareness of our mortality affects us all. *Social and Personality Psychology Compass, 2*(2), 909–928.

Arndt, J., Solomon, S., Kasser, T., & Sheldon, K. M. (2004). The urge to splurge: A terror management account of materialism and consumer behavior. *Journal of Consumer Psychology, 14*(3), 198–212.

Bauman, Z. (1991). *Modernity and its ambivalence*. Cambridge: Polity.

Bauman, Z. (1992). Survival as a social construct. *Theory, Culture and Society, 9*(1), 1–36.

Bauman, Z. (2004). *Identity*. Cambridge: Polity.

Becker, E. (1973). *The denial of death*. New York: Simon and Schuster.

Ben-Ari, O. T., Florian, V., & Mikulincer, M. (1999). The impact of mortality salience on reckless driving: A test of terror management mechanisms. *Journal of Personality and Social Psychology, 76*, 35–45.

Berger, P. (1969). *The social reality of religion*. London: Faber.

Berger, P. (1973). *The social reality of religion*. Harmondsworth: Penguin.

Berlant, L. (2012, June 15). Interview in *Rorotoko: Cutting-Edge Intellectual Interviews*. http://rorotoko.com/interview/20120605berlantlaurenoncruelop timism/?page=1. Accessed 21 Dec 2015.

Biermann, F., & Boas, I. (2008). Protecting climate refugees: The case for a global protocol. *Environment: Science and Policy for Sustainable Development, 50*(6), 8–17.

Bracken, P. J. (2001). Post-modernity and post-traumatic stress disorder. *Social Science and Medicine, 53*(6), 733–743.

Bruner, J. (1986). *Actual minds: Possible worlds*. Harvard: Harvard University Press.

Burke, B. L., Martens, A., & Faucher, E. H. (2010). Two decades of terror management theory: A meta-analysis of mortality salience research. *Personality and Social Psychology Review, 14*(2), 155–195.

Coughlan, H., Tiedt, L., Clarke, M., Kelleher, I., Tabish, J., Molloy, C., .Harley, M. and Cannon, M. (2014). Prevalence of DSM-IV mental disorders, deliberate self-harm and suicidal ideation in early adolescence: An Irish population-based study. Journal of Adolescence, 37(1), 1–9.

Cox, C. R., Goldenberg, J. L., Arndt, J., & Pyszczynski, T. (2007). Mother's milk: An existential perspective on negative reactions to breast-feeding. *Personality and Social Psychology Bulletin, 33*(1), 110–122.

Cox, C. R., Arndt, J., Pyszczynski, T., Greenberg, J., Abdollahi, A., & Solomon, S. (2008). Terror management and adults' attachment to their parents: The safe haven remains. *Journal of Personality and Social Psychology, 94*(4), 696.

Craib, I. (1998). *Experiencing identity*. London: Sage.

Croft, S. (2012). Constructing ontological insecurity: The insecuritization of Britain's Muslims. *Contemporary Security Policy, 33*(2), 219–235.

Dickinson, J. L. (2009). The people paradox: Self-esteem striving, immortality ideologies, and human response to climate change. *Ecology and Society, 14*(1), 34.

Du, H., & Jonas, E. (2015). Being modest makes you feel bad: Effects of the modesty norm and mortality salience on self-esteem in a collectivistic culture. *Scandinavian Journal of Psychology, 56*, 86–98.

Erikson, E. (1950). *Childhood and society*. New York: Norton.

Fernandez, S., Castano, E., & Singh, I. (2010). Managing death in the burning grounds of Varanasi, India: A terror management investigation. *Journal of Cross-Cultural Psychology, 41*(2), 182–194.

Frankl, V. E. (1963). *Man's search for meaning: An introduction to logotherapy*. New York: Washington Square Press.

Fritsche, I., & Häfner, K. (2012). The malicious effects of existential threat on motivation to protect the natural environment and the role of environmental identity as a moderator. *Environment and Behavior, 44*(4), 570–590.

Fritsche, I., Jonas, E., Fischer, P., Koranyi, N., Berger, N., & Fleischmann, B. (2007). Mortality salience and the desire for offspring. *Journal of Experimental Social Psychology, 43*(5), 753–762.

Fritsche, I., Jonas, E., Kayser, D. N., & Koranyi, N. (2010). Existential threat and compliance with pro environmental norms. *Journal of Environmental Psychology, 30*(1), 67–79.

Gergen, K. J. (1991). *The saturated self: Dilemmas of identity in contemporary life*. New York: Basic Books.

Giddens, A. (1990). *The consequences of modernity*. Cambridge: Polity Press.

Giddens, A. (1991). *Modernity and self-identity*. Cambridge: Polity Press.

Giddens, A. (1992). *The transformation of intimacy*. Cambridge: Polity.

Gifford, R. (2011). The dragons of inaction: Psychological barriers that limit climate change mitigation and adaptation. *American Psychologist, 66*(4), 290–302.

Goldenberg, J. L. (2005). The body stripped down: An existential account of ambivalence toward the physical body. *Current Directions in Psychological Science, 14*, 224–228.

Goldenberg, J. L., & Roberts, T. A. (2004). The beast within the beauty: An existential perspective on the objectification and condemnation of women. In J. Greenberg, S. L. Koole, & T. Pyszczynski (Eds.), *Handbook of experimental existential psychology* (pp. 71–85). New York: Guilford.

Goldenberg, J. L., Pyszczynski, T., Greenberg, J., Solomon, S., Kluck, B., & Cornwell, R. (2001). I am not an animal: Mortality salience, disgust, and the denial of human creatureliness. *Journal of Experimental Psychology: General, 130,* 427–435.

Goldenberg, J. L., Cox, C. R., Pyszczynski, T., Greenberg, J., & Solomon, S. (2002). Understanding human ambivalence about sex: The effects of stripping sex of meaning. *The Journal of Sex Research, 39*(4), 310–320.

Greenberg, J., Solomon, S., & Pyszczynski, T. (1997). Terror management theory of self-esteem and cultural worldviews: Empirical assessments and conceptual refinements. *Advances in Experimental Social Psychology, 29,* 61–139.

Griffin, J. (2010). *The lonely society?* London: Mental Health Foundation. http://www.mentalhealth.org.uk/content/assets/PDF/publications/thelonelysocietyreport.pdf. Accessed 18 Dec 2015.

Hansen, J., Winzeler, S., & Topolinski, S. (2010). When the death makes you smoke: A terror management perspective on the effectiveness of cigarette on-pack warnings. *Journal of Experimental Social Psychology, 46*(1), 226–228.

Hanson-Easey, S., Williams, S., Hansen, A., Fogarty, K., & Bi, P. (2015). Speaking of climate change: A discursive analysis of lay understandings. *Science Communication, 37*(2), 217–239.

Hayes, J., Schimel, J., Arndt, J., & Faucher, E. H. (2010). A theoretical and empirical review of the death-thought accessibility concept in terror management research. *Psychological Bulletin, 136*(5), 699–739.

Heine, S. J., Proulx, T., & Vohs, K. D. (2006). The meaning maintenance model: On the coherence of social motivations. *Personality and Social Psychology Review, 10,* 88–110.

Hollander, N. C. (2009a). When not knowing allies with destructiveness: Global warning and psychoanalytic ethical non-neutrality. *International Journal of Applied Psychoanalytic Studies, 6,* 1–11.

Hollway, W. (2016, January 19). Cited from personal correspondence with author.

Jaspal, R., Nerlich, B., & Koteyko, N. (2012). Contesting science by appealing to its norms: Readers discuss climate science in The Daily Mail. *Science Communication, 25,* 383–410.

Jessop, D. C., & Wade, J. (2008). Fear appeals and binge drinking: A terror management theory perspective. *British Journal of Health Psychology, 13*(4), 773–788.

Kashima, E. S., Halloran, M., Yuki, M., & Kashima, Y. (2004). The effects of personal and collective mortality salience on individualism: Comparing

Australians and Japanese with higher and lower self-esteem. *Journal of Experimental Social Psychology, 40*(3), 384–392.

Kasser, T., & Sheldon, K. M. (2000). Of wealth and death: Materialism, mortality salience, and consumption behavior. *Psychological Science, 11*, 348–351.

Kinnvall, C. (2004). Globalization and religious nationalism: Self, identity, and the search for ontological security. *Political Psychology, 25*(5), 741–767.

Kirkpatrick, L. A., & Navarrete, C. D. (2006). Reports of my death anxiety have been greatly exaggerated: A critique of terror management theory from an evolutionary perspective. *Psychological Inquiry, 17*(4), 288–298.

Koole, S., & Van den Berg, A. (2005). Lost in the wilderness: Terror management, action-orientation, and nature evaluation. *Journal of Personality and Social Psychology, 88*, 1014–1028.

Laing, R. D. (1960). *The divided self.* Harmondsworth: Penguin.

Last, A. (2013). Negotiating the inhuman: Bakhtin, materiality and the instrumentalization of climate change. *Theory, Culture and Society, 30*(2), 60–83.

Layard, R. (2006). The case for psychological treatment centres. *British Medical Journal, 332*, 1030–1032.

Leary, M. R., & Schreindorfer, L. S. (1997). Unresolved issues with terror management theory. *Psychological Inquiry, 8*(1), 26–29.

Lerner, M. J. (1997). What does the belief in a just world protect us from: The dread of death or the fear of understanding suffering? *Psychological Inquiry, 8*(1), 29–32.

Lifton, R. J. (1979). *The broken connection: On death and the continuity of life.* Washington, DC: American Psychiatric Press, Inc.

Lucas, C., Leith, P., & Davison, A. (2015). How climate change research undermines trust in everyday life: A review. *Wiley Interdisciplinary Reviews: Climate Change, 6*(1), 79–91.

Ma-Kellams, C., & Blascovich, J. (2011). Culturally divergent responses to mortality salience. *Psychological Science, 22*(8), 1019–1024.

Martin, I. M., & Kamins, M. A. (2010). An application of terror management theory in the design of social and health-related anti-smoking appeals. *Journal of Consumer Behaviour, 9*, 172–190.

McCabe, S., Spina, M. R., & Arndt, J. (2016). When existence is not futile: The influence of mortality salience on the longer-is-better effect. *British Journal of Social Psychology*. doi:10.1111/bjso.12143.

McGregor, I., Zanna, M. P., Holmes, J. G., & Spencer, S. J. (2001). Compensatory conviction in the face of personal uncertainty: Going to extremes and being oneself. *Journal of Personality and Social Psychology, 80*, 472–488.

McNamara, K. E., & Gibson, C. (2009). 'We do not want to leave our land': Pacific ambassadors at the United Nations resist the category of 'climate refugees'. *Geoforum, 40*(3), 475–483.

Mellor, P. A., & Shilling, S. (1993). Modernity, self-identity and the sequestration of death. *Sociology, 27*(3), 411–431.

Mikulincer, M., Florian, V., & Hirschberger, G. (2003). The existential function of close relationships: Introducing death into the science of love. *Personality and Social Psychology Review, 7*(1), 20–40.

Mnguni, P. P. (2010). Anxiety and defense in sustainability. *Psychoanalysis, Culture and Society, 15*(2), 117–135.

Muraven, M., & Baumeister, R. F. (1997). Suicide, sex terror, paralysis, and other pitfalls of reductionist self-preservation theory. *Psychological Inquiry, 8*(1), 36–40.

Navarrete, C. D., & Fessler, D. M. T. (2005). Normative bias and adaptive challenges: A relational approach to coalitional psychology and a critique of terror management theory. *Journal of Evolutionary Psychology, 3*, 297–325.

Ness, J., Hawton, K., Bergen, H., Cooper, J., Steeg, S., Kapur, N., et al. (2015). Alcohol use and misuse, self-harm and subsequent mortality: An epidemiological and longitudinal study from the multicentre study of self-harm in England. *Emergency Medicine Journal, 32*(10), 793–799.

Olausson, U. (2011). "We're the ones to blame": Citizens' representations of climate change and the role of the media. *Environmental Communication: A Journal of Nature and Culture, 5*, 281–299.

Park, C. L. (2005). Religion as a meaning-making framework in coping with life stress. *Journal of Social Issues, 61*(4), 707–729.

Pelham, B. W. (1997). Human motivation has multiple roots. *Psychological Inquiry, 8*(1), 44–47.

Pienaar, M. (2011). An eco-existential understanding of time and psychological defenses: Threats to the environment and implications for psychotherapy. *Ecopsychology, 3*(1), 25–39.

Pozzebon, M., & Pinsonneault, A. (2005). Challenges in conducting empirical work using structuration theory: Learning from IT research. *Organization Studies, 26*(9), 1353–1376.

Pyszczynski, T., Greenberg, J., Solomon, S., & Maxfield, M. (2006). On the unique psychological import of the human awareness of mortality: Theme and variations. *Psychological Inquiry, 17*, 328–356.

Routledge, C., Ostafin, B., Juhl, J., Sedikides, C., Cathey, C., & Liao, J. (2010). Adjusting to death: The effects of mortality salience and self-esteem on psychological well-being, growth motivation, and maladaptive behavior. *Journal of Personality and Social Psychology, 99*(6), 897.

Schimel, J., Simon, L., Greenberg, J., Pyszczynski, T., Solomon, S., Waxmonsky, J., & Arndt, J. (1999). Stereotypes and terror management: Evidence that mortality salience enhances stereotypic thinking and preferences. *Journal of Personality and Social Psychology, 77*(5), 905–926.

Seale, C. (2002). Resurrective practice and narrative. In M. Andrews, S. D. Sclater, C. Squire, & A. Treacher (Eds.), *Lines of narrative: Psychosocial perspectives* (pp. 36–47). London: Routledge.

Solomon, S., Greenberg, J., & Pyszczynski, T. (1991). A terror management theory of social behavior: The psychological functions of self-esteem and cultural worldviews. *Advances in experimental social psychology, 24*, 93–159.

Stoll-Kleemann, S., O'Riordan, T., & Jaeger, C. C. (2001a). The psychology of denial concerning climate mitigation measures: Evidence from Swiss focus groups. *Global Environmental Change, 11*(2), 107–117.

Threadgold, S. (2012). 'I reckon my life will be easy, but my kids will be buggered': Ambivalence in young people's positive perceptions of individual futures and their visions of environmental collapse. *Journal of Youth Studies, 15*(1), 17–32.

Vail, K. E., Juhl, J., Arndt, J., Vess, M., Routledge, C., & Rutjens, B. T. (2012). When death is good for life: Considering the positive trajectories of terror management. *Personality and Social Psychology Review, 16*(4), 303–329.

Vallacher, R. R. (1997). Grave matters. *Psychological Inquiry, 8*(1), 50–54.

Van Marle, F., & Maruna, S. (2010). 'Ontological insecurity' and 'terror management': Linking two free-floating anxieties. *Punishment and Society, 12*(1), 7–26.

Vandenberg, B. (1991). Is epistemology enough? An existential consideration of development. *American Psychologist, 46*(12), 1278–1286.

Verhaeghe, P. (2014). *What about Me?: The struggle for identity in a market-based society*. London: Scribe Publications.

Vess, M., & Arndt, J. (2008). The nature of death and the death of nature: The impact of mortality salience on environmental concern. *Journal of Research in Personality, 42*, 1376–1380.

Wakimoto, R. (2006). Mortality salience effects on modesty and relative self-effacement. *Asian Journal of Social Psychology, 9*, 176–183.

Winnicott, D. W. (1965). *The maturational process and the facilitating environment*. London: Hogarth.

Winnicott, D. W. (1974). *Playing and reality*. Harmondsworth: Penguin.

Yen, C.-L. (2013). It is our destiny to die: The effects of mortality salience and culture-priming on fatalism and karma belief. *International Journal of Psychology, 48*, 818–828.

Yip, K. S. (2004). The importance of subjective psychotic experiences: Implications on psychiatric rehabilitation of people with schizophrenia. *Psychiatric Rehabilitation Journal, 28*(1), 48–54.

Zaretsky, E. (2002). Trauma and dereification: September 11 and the problem of ontological security. *Constellations, 9*(1), 98–105.

7

Knowing and Not Knowing About Anthropogenic Ecological Crisis

Introduction

> We know it, but we cannot make ourselves believe in what we know. (Žižek 2009, p. 454)

Although to some extent it still operates from the margins of debates and research around environmental problems, psychology and behaviour, there is a growing interdisciplinary literature informed by psychodynamic concepts (e.g. Dickinson 2009; Hollander 2009; Newell and Pitman 2010; Opotow and Weiss 2000; Orr 2002; Randall 2009; Rustin 2010; Weintrobe 2012). Elements of psychoanalytic thinking are even beginning to find their way into policy debates and reports (Swim et al. 2011; Hackmann and St. Clair 2012; Norgaard 2009).[1] There is also a mounting body of empirical work that utilizes defence mechanisms as ways of categorizing and interpreting research findings in this area; if not necessarily engaging with a psychoanalytic framework more deeply (e.g. Hanson-Easey et al. 2015; Olausson 2011; Jaspal et al. 2012; Norgaard 2011; Stoll-Kleemann et al. 2001).

© The Author(s) 2016
M. Adams, *Ecological Crisis, Sustainability and the Psychosocial Subject*,
DOI 10.1057/978-1-137-35160-9_7

When Michael Rustin asks 'can a psychoanalytic perspective on [climate change] enable us to see anything we might not otherwise have noticed?' (2010, p. 475), the answer, increasingly, seems to be in the affirmative.

Psychoanalysis, as a primary example of a 'depth psychology' (Billig 1999, p. 12), is arguably well positioned to offer insights regarding the affective, irrational and hidden dimension of motivations to act (or not) in particular ways, missed by those approaches dominating the applied field. The conceptualization of defence mechanisms is a central tenet of clinical psychoanalysis, but also in theoretical, applied and social versions. It is this area that has come to the fore in the context of anthropogenic ecological crisis, often in conjunction with related concepts in fields such as existential and cognitive psychology.

Defence Mechanisms

Freud initially conceptualized 'defences' as a counterforce to the expression of unruly instinctual demands, and shifted between the use of defence mechanisms, repression and denial as a general term (e.g. Freud 1896, 1926). He understood defence mechanisms to be the varied manifestations of, in effect, the striving to keep the more secret aspects of ourselves secret *from* ourselves, not just from others (Billig 1999, p. 13). Anna Freud subsequently differentiated various types of defence mechanism (Freud 1936), and the list has been extended, modified and debated ever since (e.g. Blackman 2004). A basic list includes many terms that have slipped into popular usage, such as denial, displacement, inhibition, intellectualization, projection, splitting, reaction formation, repression and suppression (Freud 1936; Hentschel et al. 2004). Today the psychological proclivity for defence mechanisms, and the sense of what is being defended, is understood in a broad variety of ways: as a ubiquitous predisposition in the context of mortality and our awareness of it (Dickinson 2009); as part of a cognitive drive for consistency (Stoll-Kleemann et al. 2001); as a consequence of a regressive fear of dependency (Hollander 2009); or as rooted in the need to maintain a positive self-image (Cooper 1998).

Although the empirical study of defence mechanisms has long been beset by methodological criticisms (Cramer 2000), their existence remains a fundamental tenet of contemporary psychoanalysis (Akhtar 2009); and identifying and working with defence mechanisms remains a key element of clinical practice (Clark 1998; Cramer 2006). The history, aetiology, typology and shifting clinical understanding of defence mechanisms is fascinating but is explored no further here. Instead, we are concerned with how defence mechanisms have been conceptualized and applied to the relationship between human behaviour and ecological degradation.[2]

The conceptualization of defence mechanisms in the context of (un) sustainable behaviour is broadly in line with clinical definitions. It is claimed that they are triggered by situations where we are informed or warned of 'environmental' dangers and/or of the contributing role our own behaviour to those dangers (Hamilton 2013). These situations bring us face-to-face with issues relating to our dependency, mortality and responsibility. They are situations which, according to a psychodynamic perspective, generate affect—anxiety, fear, depression, anger, guilt and helplessness—that we find difficult to handle psychologically, and therefore, 'manage', through unconscious psychic dynamics (Weintrobe 2010). Considering how we cope with this affective discomfort returns us to the conceptual territory of the defence mechanism.

Defence Mechanisms in the Anthropocene

Clear evidence that concepts of defence and denial have made it into the mainstream problematization of the 'information-deficit model' of climate (in)action is their appearance in Intergovernmental Panel on Climate Change discourse. Their reference to defence mechanisms in the fifth Assessment Report (AR5) was as follows:

> There is evidence of cognitive dissonance and strategic behaviour in both mitigation and adaptation. Denial mechanisms that overrate the costs of changing lifestyles, blame others, and that cast doubt on the effectiveness of individual action or the soundness of scientific knowledge are well docu-

mented… as is the concerted effort by opponents of climate action to seed and amplify those doubts. (Fleurbaey et al. 2014, p. 300)

Fleurbaey et al.'s comments provide a reasonable summary of the 'official' status of defence mechanisms. The authors identify two aspects of denial. The final sentence of the quote refers to avowed 'opponents of climate action' who facilitate doubt and uncertainty towards the soundness of the science involved in climate science in particular, and/or our ability to do anything about it. Denial*ism* is often considered a more appropriate term here, because it is the *evidence* that is denied, via the use of various strategies, and in line with specific beliefs and worldviews and/or vested interests (Diethelm and McKee 2009; Dunlap 2013; Oreskes and Conway 2010). There is now an ample literature documenting prevalent forums and narratives of climate denialism (see Dunlap 2013 for an overview).

The main focus in this chapter are these more mundane forms of denial, sketched in broad cognitive terms by Fleurbaey et al., along with variations of defence mechanisms that have been identified in the literature. However, it is important to recognize the interrelationship of these different forms of denial. The concerted efforts of the denialism industry can also seed and amplify uncertainty and scepticism in more general populations, making its discourse, and its very presence, prime candidates for active recruitment into 'ordinary' everyday processes of denial.

At their simplest, defence mechanisms allow us to 'deny or pretend the problem is not there, or that it is the responsibility of someone else' (Lertzman 2008, p. 16). As this definition suggests, there is a tendency to equate defence mechanisms *with* denial, although clinical literature has normally considered denial as a specific variation of them.[3] In terms of anthropogenic ecological crisis, although it is studied in various guises, denial is far and away the most commonly discussed defence mechanism (e.g. Hamilton 2012; Jacques et al. 2008; Norgaard 2011; Opotow and Weiss 2000). This is partly because it can be incorporated into analysis and discussion without having to engage with the detail of the psychoanalytic paradigm. We all seem to *know* what denial means, so it can be sprung on us without too much concern or need for clarification.

Ways of Knowing and Not Knowing at the Same Time

Denial also has an undoubted conceptual utility however, not least in building on Stanley Cohen's remarkable work on the human capacity for denial in the context of human suffering and atrocities (Cohen 1996, 2001). Cohen only briefly applied his conceptualization of denial to climate change (Cohen 2013). However, numerous others have taken up his framework or developed overlapping ones (e.g. Opotow and Weiss 2000; Lertzman 2015; Lorenzoni et al. 2007; Norgaard 2006a, b; Washington and Cook 2011). His typology can help us clarify as to what is being denied and how, when we talk about the denial of anthropogenic climate change and related forms of ecological degradation.

Cohen offers a composite definition of denial as 'an unwillingness to accept the reality of uncomfortable, painful facts (and/or unconsciously) the repression of such facts' (2013, pp. 72–3). He elaborates further that denial really only applies when 'the difference between knowing and not knowing is not what it seems to be…' (p. 73). Rather than outright lying or inadvertent ignorance, denial involves 'a state of knowing and not knowing at the same time' (p. 73). This knowing/non-knowing relationship can take different forms, and here Cohen makes a categorical distinction between literal, interpretive and implicatory denial (Cohen 2001).

Literal Denial

Literal denial is the denial that something exists or has happened at all. Literal denial of climate change involves rejecting the scientific basis— 'the evidence does not point to global warming at all'. Analysis of self-defined climate skeptic outputs indicates that is by far the most prevalent basis for rejecting climate change as an urgent social problem. As such, it lends support for Cohen's claim that this is 'the strongest and most primitive form of denial' (2013, p. 73). Explanations of why there is an *apparent* consensus are often conspiracy oriented—i.e. a left-liberal power block as a cover for increasing state power and/or for pushing their own policy agendas (Dunlap and McCright 2010; Lewandowsky et al. 2013). They need not be fully formed conspiracy theories to work effectively as forms of literal denial (e.g. 'I've heard the science is exaggerated'), as long as they head off the potential for anxiety and insecurity. Rationalizations such as this draw on strategies related to other defence mechanisms, and more is said about them below. There is no reason why literal denial cannot be seen to serve this function at an individual, interpersonal and broader social level. Conversations, media reporting, blogs, twitter feeds are all forms of communication that might circulate and reinforce literal denial. Explicitly political 'denialism', described above, is significant in this context. It is an explicit attempt to 'seed and amplify' mundane forms of literal denial.

Elsewhere something akin to literal denial is given a slightly different inflection. Based on her ethnographic work, Norgaard sees denial as working through a more active and everyday process of 'selective attention' (Norgaard 2006a). Selective attention, Norgaard argues, is aimed at managing feelings of fear and helplessness via deciding what to and what not to think about. Practical strategies include 'screening out—for example—painful information about problems for which one does not have solutions (e.g. "I don't really know what to do, so I just don't think about that")' (2006a, p. 385); focusing on small behaviour changes, or trying to avoid thinking about the future. These strategies imply a more active process of denial at some level in terms of a subject 'controlling' or 'restricting' their own exposure to information (Hamilton 2012, p. 19). It is not a 'genuine inadvertent ignorance' which would exclude it from the category of denial in Cohen's account, nor is it a literal rejection of the science. It is perhaps a more 'artful' and 'perverse' rejection of attending to, noticing of, the science base, rather than a noticing-then-rejecting (Hoggett 2012, p. 60). It might still be classed as a form of simultaneous 'knowing and not-knowing' and a variation of literal denial.

Interpretive Denial

Interpretive denial involves denying conventional interpretations of a phenomenon, without refuting its existence per se. Interpretive denial might involve the acceptance that climate change is happening, while denying that it is anthropogenic. Associated explanations may draw on ideas about natural fluctuations, the inability of human interventions to address or reverse warming, or an emphasis on the supposed benefits of anthropogenic climate change for human societies, other species or landscapes. This form of denial is most often associated with the 'euphemisms, technical jargon and word-changing' of governments and big business (Norgaard 2011, p. 10; Washington and Cook 2011). However, it also informs the denialism discourses of avowed 'climate sceptics'. Attempts to positively spin climate change, or refute its anthropogenic origins, are well-documented tenets of the armoury of climate 'skepticism' (e.g. McCright and Dunlap 2000; Oreskes 2010; Washington and Cook 2011). Everyday examples of mundane denial similarly though less systematically draw on similar rationalizations, e.g. 'what about less influenza virus for example if it's warmer weather, I don't know' (Hobson and Niemeyer 2013, p. 408). Interpretive denial involves a selective acceptance of climate science, in which humans are exonerated of blame or consequences muted; hence there is little credence given to the need to act collectively.

Implicatory Denial

Implicatory denial requires denying the significance of the reality of a thing or event, without refuting its existence, or, necessarily, standard interpretations of it (Cohen 2013). Although social and applied psychoanalysis does not consistently use terminology in the same way, for our purposes implicatory denial is more or less interchangeable with 'disavowal'; whereby 'reality is more accepted, but its significance is minimized' (Weintrobe 2013, p. 7). This reflects Freud's own understanding, where, as Hollander notes, 'disavowal did not erase the threatening idea or perception so much as it did its *meaning to the subject*' (Hollander 2009, p. 3; emphasis added). Disavowal demands the denial of 'emotional, moral and political' implications of climate change (Cohen 2013, p. 73). It is in this category that we might expect to see a fuller acceptance of climate science, but a selective interpretation of what kind of action is needed:

harm or risk of harm is recognized, but not seen as psychologically disturbing or as carrying a moral imperative to act... Unlike literal or interpretive denial, knowledge itself is not at issue, but doing the 'right' thing with the knowledge. These are issues of mobilization, commitment and involvement. (Cohen 2001, p. 9)

Constructing ecological crisis as a technical problem is a straightforward example of disavowal—it marginalizes understandings that require changes in lifestyle, social structure, political processes—because they are experienced as more threatening, according to the logic of defence mechanisms. 'Techno-fix' solutions, for example, implicate scientists, engineers, rather than politicians, the public sphere, civil society (Klein 2014).

Implicatory denial is often described in relation to the defence mechanisms of splitting and projection.[4] At a key moment in the narrative of Squarzoni's 'journey through climate science' he ponders the complex contradictions that can beset good intentions and engender a 'split personality':

> We're caught in so many contradictions. Page left: we know we're headed for a wall. Page right: we go on living in a fantasy land where there's no contradiction between our material desires and preserving the planet. We know, but we don't make changes. Our initial ignorance has been replaced by some sort of split personality. To keep living in this fantasy world we play hide-and-seek with what we know. In our delusion we feel the urgency to act... without believing we have the means to do so. We know another story has begun but we continue to act like it's nothing. And the worse thing is... it feels pretty good. (Squarzoni 2014, pp. 242–243)

Despite its origins in complex conceptual models of object relations theory, the notion of splitting has been translated loosely and variedly in its application to the psychodynamics of responding to anthropogenic ecological crisis—be it in terms of the idealization of other cultures or historical periods (Weintrobe 2010); the splitting of losses into distant futures or places (Randall 2009); or nuances in forms of climate change denial (Leviston and Walker 2012). What most definitions share is an understanding that splitting is the basis for defending against the anxiety induced by an anthropogenic ecological crisis.

The mechanics of splitting, often involve descriptions of how 'internal objects' – unconscious thought and emotion—are disavowed, split off and *projected* onto 'external objects'—other people, ideas and groups. In other words, denial creates a psychic splitting, and projection places the split-off element 'outside'. Psychodynamics are considered to involve more complex interactions than this linear explanation suggests, but for our purposes, it does at least suggest a relatively straightforward process. If we take, broadly speaking, 'fear of loss' as an unconscionable emotion arising from anthropogenic ecological crisis communication, the tendency to consider problems as being far in the future involves denial, splitting and projection—the fear of loss is denied, split off from a 'safe' present (which is soothed with 'small steps' or 'techno-fix' solutions), and projected into a distant (and therefore 'safe') future (Randall 2009, p. 119; Opotow and Weiss 2000; Hamilton 2013, Dickinson 2009).

Such solutions might overlap, or may be distinct from, the dynamics of avoiding the implications for oneself, one's immediate ingroup, or the groups we identify with collectively such as class, generation or nation—'they are more implicated than I am'. This might involve a denial of how likely 'we' are to be affected by the fallout of ecological crisis, or a denial of the extent to which we are implicated as responsible for it.[5] Norgaard found evidence of implicatory denial in her ethnographic and interview studies involving inhabitants of a small Norwegian town (2006a, b; 2011). She observes 'the use of stock stories to frame potentially disturbing information about climate change in a more positive light' (2006a, p. 358). They include idealized depictions of the Norwegian national identity, 'that, in emphasizing simplicity, purity, and innocence, deflects attention from the fact that Norwegian wealth, political economy, and way of life are intimately connected to the problem of global warming' (2006a, p. 359); as well as minimizing the responsibility of one's self or nation by emphasizing the greater responsibility of others (e.g. Americans).[6]

Forms of literal denial, and to a lesser extent interpretive denial, are the primary focus of sociologists concerned with the strategies and practices of climate denialism, introduced above (e.g. Dunlap 2013; Jacques et al.

2008). Implicatory and interpretive denial might now be understood as general terms for a multitude of mundane and everyday defences utilized to deflect the magnitude and responsibility of anthropogenic ecological crisis—the 'rich, convoluted and ever-increasing vocabulary for bridging the moral and psychic gap between what you know and what you do' (Cohen 2001, p. 9). However, this vocabulary is not 'merely' psychological; it emerges, is tested and validated (or not) in particular places at particular times. It relies on sociocultural, interpersonal and psychological channels for meaningful dissemination. In fact, as these channels are integrated and simultaneous, we can refer to the vocabularies of denial as *psychosocial.*

In the previous chapter, the concepts of ontological insecurity and mortality salience were utilized to frame ecological crisis as an embodied and affective threat. In this chapter, it has been suggested that our attempts to process this threat can set in motion defence mechanisms that are made visible through particular ways of coping that can be both adaptive and maladaptive. The next step, in Chapter 8, is to identify how these discernable signs of defence mechanisms are manifest *socially*; and might be best thought of, therefore, as psychosocial mechanisms.

Notes

1. According to Hackmann and St. Clair (2012, p. 19), one of the 'transformative cornerstones' of social science research agenda is interpretation and subjective sense making, which 'he personal and collective values, beliefs, assumptions, interests, worldviews, hopes, needs and desires that underlie people's experiences of and responses – or lack of responses – to processes of global change'. Questions arising from this cornerstone, Hackmann and St. Clair assert, include those concerned with 'scepticism and denialism in the face of potentially cataclysmic processes of climate change' (2012, p. 19).
2. For the most part, recent work does not engage with the complex detail of the psychoanalytic theory in which the concept of defence mechanisms originated. The specific types of anxiety that supposedly

give rise to various mechanisms and their relationship to a psychoanalytic model of self, developed, for example, in Anna Freud's work (Freud 1936), are rarely revisited. Similarly, most accounts focus on a handful of mechanisms, namely denial, disavowal, projection and splitting.

3. Anna Freud described it as the 'simplest defence mechanism' (1936, p. 93); a necessarily infantile form of defence subsequently abandoned for more complex operations. In her attempt to characterize defence mechanisms in a hierarchical form, from least to most complex, Cramer similarly locates denial as a low-level defence, established early in life, and successfully adaptive at this stage, but used minimally or as a maladaptive response in later life (Cramer 1987; Cramer and Block 1998). Though this portrayal skates over numerous complexities and contingencies (Cramer 2000, p. 643), there does appear to be consensus that denial is a primary and formative defence mechanism.

4. The concept of splitting is indebted to Melanie Klein's work on early child development (e.g. Klein 1946, 1952, 1957); and elaborated in object relations theory (e.g. Fairbairn 1949; Kernberg 1976; Greenberg 1983; Guntrip 1992).

5. In both cases there are clear cases to be made where this is literally true – the developed North is more responsible for climate change than the developing or underdeveloped South; whilst, on the whole, the latter is the most exposed to those consequences – the infamous 'double injustice' of climate change (Gough 2011).

6. How sociocultural contexts, including dominant narratives, shape denial is the focus of Chapter 8.

References

Akhtar, S. (2009). *Comprehensive dictionary of psychoanalysis*. London: Karnac Books.

Billig, M. (1999). Commodity fetishism and repression: Reflections on Marx, Freud and the psychology of consumer capitalism. *Theory & Psychology, 9*(3), 313–329.

Blackman, J. (2004). *101 defences: How the mind shields itself.* Hove: Psychology Press.

Clark, A. J. (1998). *Defence mechanisms in the counseling process.* London: Sage.

Cohen, S. (1996). Government responses to human rights reports: Claims, denials, and counterclaims. *Human Rights Quarterly, 18*(3), 517–543.

Cohen, S. (2001). *States of denial: Knowing about atrocities and suffering.* New York: Wiley.

Cohen, S. (2013). Discussion. In S. Weintrobe (Ed.), *Engaging with climate change: Psychoanalytic and interdiscipinary perspectives* (pp. 72–79). London: Routledge.

Cooper, S. H. (1998). Changing notions of defence within psychoanalytic theory. *Journal of Personality, 66*(6), 947–964.

Cramer, P. (1987). The development of defense mechanisms. *Journal of Personality, 55*(4), 597–614.

Cramer, P. (2000). Defense mechanisms in psychology today: Further processes for adaptation. *American Psychologist, 55*(6), 637.

Cramer, P. (2006). *Protecting the self: Defence mechanisms in action.* New York: Guilford Press.

Cramer, P., & Block, J. (1998). Preschool antecedents of defense mechanism use in young adults: A longitudinal study. *Journal of Personality and Social Psychology, 74*(1), 159.

Dickinson, J. L. (2009b). The people paradox: Self-esteem striving, immortality ideologies, and human response to climate change. *Ecology and Society, 14*(1), 34.

Diethelm, P., & McKee, M. (2009). Denialism: What is it and how should scientists respond? *The European Journal of Public Health, 19*(1), 2–4.

Dunlap, R. E. (2013). Climate change skepticism and denial: An introduction. *American Behavioral Scientist, 57*(6), 691–698.

Dunlap, R. E., & McCright, A. M. (2010). Climate change denial: Sources, actors and strategies. In C. Lever-Tracy (Ed.), *Routledge handbook of climate change and society* (pp. 240–259). London: Routledge.

Fairbairn, W. R. D. (1949). Steps in the development of an object-relations theory of the personality. *British Journal of Medical Psychology, 22*(1–2), 26–31.

Fleurbaey, M., Kartha, S., Bolwig, S., Chee, Y. L., Chen, Y., Corbera, E., Lecocq, F., Lutz, W., Muylaert, M. S., Norgaard, R. B., Okereke, C., & Sagar, A. D. (2014). Sustainable development and equity. In O. Edenhofer, R. Pichs-Madruga, Y. Sokona, E. Farahani, S. Kadner, K. Seyboth, A. Adler, I. Baum,

S. Brunner, P. Eickemeier, B. Kriemann, J. Savolainen, S. Schlömer, C. von Stechow, T. Zwickel, & J. C. Minx (Eds.), *Climate change 2014: Mitigation of climate change. Contribution of Working Group III to the fifth assessment report of the Intergovernmental Panel on Climate Change.* Cambridge: Cambridge University Press.

Freud, A. (1936) [1968]. *The ego and the mechanisms of defence* (trans: Baines, C.). London: Hogarth Press.

Freud, S. (1896). Further remarks on the neuro-psychoses of defence (trans: Strachey, J.). In *The standard edition of the complete works of Sigmund Freud*, vol. 3. (1968) (pp. 162–85). London: Hogarth.

Freud, S. (1926). The question of lay analysis (trans: Strachey, J.). In *Penguin Freud library, vol. 15, Historical and expository works*. Harmondsworth: Penguin, 1986.

Gough, I. (2011). *Climate change, double injustice and social policy: A case study of the United Kingdom.* Geneva: UNRISD.

Greenberg, J. (1983). *Object relations in psychoanalytic theory.* Cambridge: Harvard University Press.

Guntrip, H. (1992). *Schizoid phenomena, object relations and the self.* New York: Karnac Books.

Hackmann, H., & St. Clair, A. L. (2012). *Transformative cornerstones of social science research for global change.* (International Social Science Council). http://www.worldsocialscience.org/documents/transformative-cornerstones. pdf. Accessed 18 Dec 2015.

Hamilton, C. (2012). What history can teach us about climate change denial. In S. Weintrobe (Ed.), *Engaging with climate change: Psychoanalytic & interdisciplinary perspectives* (pp. 16–32). London: Routledge.

Hamilton, C. (2013). What history can teach us about climate change denial. In S. Weintrobe (Ed.), *Engaging with climate change: Psychoanalytic & interdisciplinary perspectives* (pp. 16–32). London: Routledge.

Hanson-Easey, S., Williams, S., Hansen, A., Fogarty, K., & Bi, P. (2015). Speaking of climate change: A discursive analysis of lay understandings. *Science Communication, 37*(2), 217–239.

Hobson, K., & Niemeyer, S. (2013). "What sceptics believe": The effects of information and deliberation on climate change skepticism. *Public Understanding of Science, 22*(4), 396–412.

Hoggett, P. (2012). Climate change in a perverse culture. In S. Weintrobe (Ed.), *Engaging with climate change: Psychoanalytic and interdisciplinary perspectives* (pp. 56–71). London: Routledge.

Hollander, N. C. (2009). When not knowing allies with destructiveness: Global warning and psychoanalytic ethical non-neutrality. *International Journal of Applied Psychoanalytic Studies, 6*(1), 1–11.

Jacques, P., Dunlap, R. E., & Freeman, M. (2008). The organization of denial: Conservative think tanks and environmental scepticism. *Environmental Politics, 17*, 349–385.

Jaspal, R., Nerlich, B., & Koteyko, N. (2012). Contesting science by appealing to its norms: Readers discuss climate science in The Daily Mail. *Science Communication, 25*, 383–410.

Kernberg, O. (1976). *Object relations theory and clinical psychoanalysis*. New York: Jason Aronson.

Klein, M. (1946a). Notes on some schizoid mechanisms. *International Journal of Psychoanalysis, 27*, 99–110.

Klein, M. (1952). The origins of transference. *International Journal of Psychoanalysis, 33*, 433–438.

Klein, M. (1957). *Envy and gratitude. A study of unconscious sources*. New York: Basic Books.

Klein, N. (2014). *This changes everything. Capital vs the climate*. Harmondsworth: Penguin.

Lertzman, R. (2008, June 9). The myth of apathy. *The Ecologist*. http://www.theecologist.org/blogs_and_comments/commentators/other_comments/269433/the_myth_of_apathy.html. Accessed 18 Dec 2015.

Lertzman, R. (2015). *Environmental melancholia: Psychoanalytic dimensions of engagement*. London: Routledge.

Leviston, Z., & Walker, I. (2012). Beliefs and denials about climate change: An Australian perspective. *Ecopsychology, 4*(4), 277–285.

Lewandowsky, S., Oberauer, K., & Gignac, G. E. (2013). NASA faked the moon landing – Therefore, (climate) science is a hoax an anatomy of the motivated rejection of science. *Psychological Science, 24*(5), 622–633.

Lorenzoni, I., Nicholson-Cole, S., & Whitmarsh, L. (2007). Barriers perceived to engaging with climate change among the UK public and their policy implications. *Global Environmental Change, 17*(3), 445–459.

McCright, A. M., & Dunlap, R. E. (2000). Challenging global warming as a social problem: An analysis of the conservative movement's counterclaims. *Social Problems, 47*(4), 499–522.

Newell, B. R., & Pitman, A. J. (2010). The psychology of global warming: Improving the fit between the science and the message. *Bulletin of the American Meteorological Society, 91*(8), 1003–1014.

Norgaard, K. M. (2006a). "People want to protect themselves a little bit": Emotions, denial, and social movement nonparticipation. *Sociological Inquiry, 76*(3), 372–396.

Norgaard, K. M. (2006b). "We don't really want to know" environmental justice and socially organized denial of global warming in Norway. *Organization & Environment, 19*(3), 347–370.

Norgaard, K. M. (2011). *Living in denial: Climate change, emotions and everyday life*. Cambridge: MIT Press.

Norgaard, K. M. (2012). Climate denial and the construction of innocence: Reproducing transnational environmental privilege in the face of climate change. *Race, Gender & Class, 19*, 80–103.

Olausson, U. (2011). "We're the ones to blame": Citizens' representations of climate change and the role of the media. *Environmental Communication: A Journal of Nature and Culture, 5*, 281–299.

Opotow, S., & Weiss, L. (2000). New ways of thinking about environmentalism: Denial and the process of moral exclusion in environmental conflict. *Journal of Social Issues, 56*, 475–490.

Oreskes, N. (2010). My facts are better than your facts: Spreading good news about global warming. In P. Howlett & M. S. Morgan (Eds.), *How well do facts travel? The dissemination of reliable knowledge* (pp. 135–166). Cambridge: Cambridge University Press.

Oreskes, N., & Conway, E. M. (2010). *Merchants of doubt: How a handful of scientists obscured the truth on issues from tobacco smoke to global warming*. New York: Bloomsbury Publishing USA.

Orr, D. W. (2002). Four challenges of sustainability. *Conservation Biology, 16*, 1457–1460.

Randall, R. (2009). Loss and climate change: The cost of parallel narratives. *Ecopsychology, 3*, 118–129.

Rustin, M. (2010). Looking for the unexpected: Psychoanalytic understanding and politics. *British Journal of Psychotherapy, 26*(4), 472–479.

Squarzoni, P. (2014). *Climate changed: A personal journey through the science*. New York: Abrams ComicArt.

Stoll-Kleemann, S., O'Riordan, T., & Jaeger, C. C. (2001b). The psychology of denial concerning climate mitigation measures: Evidence from Swiss focus groups. *Global Environmental Change, 11*(2), 107–117.

Swim, J. K., Stern, P. C., Doherty, T. J., Clayton, S., Reser, J. P., Weber, E. U., et al. (2011). Psychology's contributions to understanding and addressing global climate change. *American Psychologist, 66*(4), 241.

Washington, H., & Cook, J. (2011). *Climate change denial: Heads in the sand.* Abingdon: Earthscan.

Weintrobe, S. (2010a). Engaging with climate change means engaging with our human nature. *Ecopsychology, 2*(2), 119–120.

Weintrobe, S. (Ed.). (2012). *Engaging with climate change: Psychoanalytic and interdisciplinary perspectives.* London: Routledge.

Žižek, S. (2009). *In defense of lost causes.* London: Verso.

8

Building a Movement Against Ourselves? Socially Organized Defence Mechanisms

Introduction

> Since all of us are in some way the beneficiaries of cheap fossil fuel, tackling climate change has been like trying to build a movement against yourself—it's as if the gay-rights movement had to be constructed entirely from evangelical preachers, or the abolition movement from slaveholders. (McKibben 2012)

Although some of the ramifications of the defence mechanism typology advanced in the previous chapter at least imply otherwise, up to this point denial has broadly been described as a psychological process. However, a growing body of literature engages with disavowal, splitting and projection as *culturally and socially organized* mechanisms (Norgaard 2011), in ways that might encourage inaction in the context of anthropogenic ecological crisis (Hollander 2009). In developing a psychosocial perspective, it is work in this area that holds the most promise and is the main focus of this chapter.

© The Author(s) 2016   **145**
M. Adams, *Ecological Crisis, Sustainability and the Psychosocial Subject*,
DOI 10.1057/978-1-137-35160-9_8

After Cohen

Again the touchstone is Cohen's sociological analysis of denial and suffering. He asserts that denial can be 'individual, psychological, personal and private – or shared social, collective and organized' (2001, p. 9). We follow Cohen's later caveat—that it is necessary to avoid bestowing 'a spurious strength and psychological depth to certain positions and people' in using 'denial' as an explanatory category (2013, p. 74). In advancing a psychosocial understanding of defence and denial, it is a fallacy to posit psychological mechanisms as discreet entities distinct from emotion, action, education, socialization (Cohen 2013, p. 76). All intersect in a 'flowing, fluid narrative' that encompasses everyday experiences and concerns, and our shared but idiosyncratic access to the continuous mediation of evidence, claims, counter-claims (2013, p. 77).

At the broadest collective level appropriate to anthropogenic ecological crisis, Cohen says this:

> Whole societies may slip into collective modes of denial not dependent on a fully-fledged Stalinist or Orwellian form of thought control. Without being told what to think about (or what not to think about) and without being 'punished' for knowing the wrong things, societies arrive at unwritten agreements about what can be publicly remembered and acknowledged. (2001, pp. 10–11)

Sociocultural forms of denial combine with personal, internal dynamics and discourses, 'official' forms—'public, collective, highly organized', often resourced, spread and maintained by the state—as well as 'microcultures' of denial within a range of institutions and organizations, from families to bureaucracies: 'the group censors itself, learns to keep silent about matters whose open discussion would threaten its self-image… organizations depend on forms of concerted ignorance, different levels of the system keeping themselves uninformed about what is happening elsewhere' (2001, p. 11). Cohen's latter description could equally apply to a psychodynamic perspective on the psychological 'system'. The basic dynamics of denial appear to be equally at play within, without and between.

Although Cohen briefly addressed climate change in relatio
sociocultural approach to denial (Cohen 2013), it is others whc ɪɪᴀᴠᴇ
begun to take up the challenge of how societies 'arrive at unwritten
agreements' about the normative implications of anthropogenic ecological crisis for human activity. Although originally a psychological
concept, defence mechanisms are increasingly understood as social psychological and socio-cultural phenomenon, specifically in relation to
human responses to ecological degradation and climate change (e.g.
Lertzman 2015; Randall 2009; Hollander 2009; Keene 2013; Rustin
2012; Norgaard 2011). Intriguingly, there is significant conceptual
overlap with sociological work focusing on ontological insecurity and
social psychological accounts of mortality salience—even if their relationship to everyday life in the Anthropocene is as yet underdeveloped.
In this chapter, I attempt to make links across these related authors,
ideas and concepts, in developing an avowedly psychosocial perspective, rather than convey a specific author's perspective or contribution.
In doing so, we return to the vocabulary of disavowal, splitting and
projection.

Bifurcation of nature

Disavowal, Splitting and Projection

The bifurcation of humans and the rest of nature might be understood
as the 'first' split (Jordan 2009; Mishan 1996; Searles 1960). Splitting
can here be interpreted as derived from a fear of vulnerability and dependency that underpins the relationship between human beings and the
earth, and between life and death; akin to the 'cosmic fear' described
earlier (Randall 2005; Searles 1972). Terror Management Theory links
these relationships in claiming that:

> individuals are motivated to distinguish themselves from the rest of nature
> because doing so facilitates the denial of human mortality… While faith in
> a transcendent reality and feelings of self-worth appear to help manage the
> awareness of death, they can do nothing to change the fact that, like all
> other plants and animals, humans die. (Vess and Arndt 2008, p. 1377)

This fear is split off in the intersecting worldviews of patriarchy, colonialism, neoliberalism and fundamentalist individualism, and projected onto denigrated others (the 'weaker' sex, the impoverished, the savage). Omnipotence and idealization are deeply inscribed in the ascendance of an industrial and scientific paradigm, animated by 'ideas of a knowable and controllable earth' (Klein 2014, p. 170), and the 'pitting of omnipotent narcissistic versions of autonomy against degraded narcissistic versions of dependence' (Layton 2014, p. 164). Such dynamics reflect a key tenet of the object relations tradition in psychoanalysis, namely 'that excessive reliance on the self arises from perceived or fantasied failures of relationship' (Rustin 2014, p. 151).[1] It is difficult, perhaps impossible, to ascertain the extent to which this 'primary' split underpins or informs other defence mechanisms. What we *can* do is consider other socially organized defence dynamics and their potential relationship to the bifurcation of nature.

Proximal and Distal Defences

Pyszczynski et al.'s dual-process of model of terror management (1999), differentiated in terms of types of denial, provides a useful framework for further consideration of defence mechanisms, as it is broad enough to incorporate the variety in the relevant literature, and elaborated upon illustratively in relation to climate change by Dickinson (2009). The model proposes that there are two types of defence—proximal and distal—that together comprise the anxiety buffer system that holds death thoughts at bay. Distal defences are tasked with corralling death thoughts in the unconscious, hence the spatial metaphor—they are kept 'far away'. They are experiential and culturally embedded and do not involve conscious engagement with the problem of death. Although more sharply focused on death-anxiety, there are close parallels between the function of distal defences and the concept of ontological security. Hence, distal defences 'entail maintaining self-esteem and faith in one's cultural worldview [and] function to control the potential for anxiety that results from knowing that death is inevitable' (Pyszczynski et al. 1999, p. 835).

The particular strategies of distal defences are counter-intuitive in that they are not logically related to the fact of one's own death. Instead, they reflect the idea, derived from Becker, that unconscious attempts to transcend death are apparent in activity that seeks to secure symbolic immortality. Such activities involve validating the ingroup and worldview one identifies with:

> Prescribing harsher punishment for moral transgressors, becoming more hostile toward outgroups, and exaggerating the extent of social consensus for one's attitudes... From a [Terror Management Theory] perspective, such defensive reactions shield individuals from fears surrounding death by enabling them to view themselves as valuable members of an eternal cultural reality that persists beyond the point of their own physical death. (Pyszczynski et al. 1999, p. 839)

The more death thoughts become 'accessible', the more these defences are relied upon. However, when death thoughts are accessible to the extent that they breach consciousness—'nearby', we turn to proximal defences. Specific proximal defences involve biased and distorted rationalizations activated by thoughts of death. They include relegating the problem of death to the distant future 'by denying one's vulnerability' (Pyszczynski et al. 1999, p. 835).

Climate Change and the Dual Process Model

Dickinson uses this model to conceptualize the barriers to addressing global climate change (Dickinson 2009; see Table 8.1). Following the logic of Terror Management Theory, denial is here understood not as a denial of ecological crisis and the anxieties they give rise to, but in terms of something more psychologically, or even biologically, fundamental that ecological crisis is a manifestation of—a denial of our own mortality. Dickinson speculates that defences are triggered by communication about ecological crises. In line with Terror Management Theory these defences *can* be productive or a hindrance for forms of behaviour—generally they encourage attempts to transcend death symbolically in

Table 8.1 Defending against anthropogenic ecological crisis (Adapted from Dickinson 2009)

Distal defences	1	transference idealization in the form of blind following and a reduction in the rational criticism of public figures, particularly charismatic leaders
	2	increased striving for self-esteem, which in Western society could mean counterintuitive increases in status-driven consumerism, materialism, and other behaviours that increase carbon emissions
	3	increased outgroup antagonism, not just between environmentalists and anti-environmentalists, but among religious groups, gangs, and other ideological groups
	4	a tendency to bolster the existing world view even if it is not sustainable
Proximal defences	1	denial of climate change, i.e. climate skeptics
	2	denial that humans are the cause of climate change
	3	a tendency to minimize or project the impacts of climate change far into the future, where they no longer represent a personal danger

various ways. In terms of ecological crisis, the danger lies in the following counter-intuitive paradox: 'the very things that bring us symbolic immortality often conflict with our prospects for survival' (2009, p. 1).

Dickinson's hypothesized proximal responses closely parallel existing theoretical and empirical work on denial and defence mechanisms in the context of ecological crisis, so there is plenty of support for this aspect of the dual-process model. Proximal defences—denying that climate change exists, and denying that humans are the cause, are analogous to Cohen's typology of literal and interpretive denial, described above. The tendency to project into the future is again echoed in Cohen's typology, this time in terms of implicatory denial or disavowal, as we have seen. Proximal defences are 'attempts to defuse a threat at roughly the same level of abstraction at which it is construed' (Pyszczynski et al. 1999, p. 839).

Dickinson argues that when climate change is presented (and experienced) as an abstract problem, proximal defences are the most likely. Distal defences are more intriguing, because, according to Dickinson, 'as conditions worsen and it becomes increasingly difficult to deny the effects of global climate

change, more people will probably switch over to distal defenses' (Dickin~~~
2009, p. 38). The closer ecological catastrophes appear, the more the dread of
annihilation becomes salient, the more desperately we try to push the dread
away, and attempt to preserve a 'sense of invulnerability' (Jordan 2009, p. 29).
We will be less willing to contemplate threats emanating from a changing
climate in consciousness, so make more unconscious effort to hold them at
bay. Dickinson refers to four possible manifestations of distal defences, which
will now be considered in more detail with reference to the wider literature.

Transference Idealization

Transference idealization is the first of Dickinson's manifestations of distal defences. It refers to the 'projection of power and importance onto
some idealized other... [when] we repress thoughts of death and project a
power and importance onto something larger that will save us' (Dickinson
2009, p. 37). 'Others' might include charismatic and/or political leaders,
as in Rustin's description of investing hope in the 'the coming of a phan-
tasied messiah' in this context (2010, p. 478). However, it is equally possible that power is projected on to multiple, contingent idealized others,
that might include celebrities, film stars, teachers or partners (Boykoff
and Goodman 2009; Pienaar 2011).

There is theoretical support for this form of transference from various
sources. Piennar talks of faith in 'the ultimate rescuer', who intercedes between
the individual and death, as a form of defence applicable to awareness of ecological crisis (see also Langford 2002). Following Yalom (1980) and Firestone
(1994), the belief that we will be rescued at the last minute requires idealizing
or merging with the idealized other. For Firestone, this defence is culturally
patterned, a specific manifestation of a broader human tendency to 'conspire
with one another to create cultural imperatives and institutions that deny the
fact of mortality' (Firestone, cited in Pienaar 2011, p. 31).

Empirical work offers some support for the concept of transference
idealization, including Lorenzoni et al.'s large mixed-methods studies of
perceived barriers to engaging with climate change in the UK population
(2007). The authors specifically highlighted 'technology will save us' as
a 'major' individual barrier to engagement with climate change issues

(2007, p. 452); echoing similar findings of earlier focus groups studies (Stoll-Kleemann et al. 2001; Blake 1999). In presenting faith in technology as an individual barrier (e.g. Gifford 2011), there is a tendency to marginalize the extent to which 'psychological' defences are coeval with social and cultural framings of problems and solutions. Randall uses a different understanding of death, loss and grief to model how we respond to climate change defensively (particularly Worden's typology of the task involved in grief and mourning), but her analysis is directly relevant here. She is explicit about the tendency to idealize others as a culturally validated 'solution narrative' that defends against the anxiety of a climate crisis. She highlights a number of narratives prevalent in the supportive group sessions she devised as spaces to have conversations about climate change (Randall 2009). She describes a 'technology will save us' narrative as one in which 'the boffins (all those professors in white coats, pebble-glasses, and crazy hair) have the answers—whether it is renewables, nuclear power, or geoengineering. Once they have their hands on the resources they need, they will deliver a world much like the one we know' (Randall 2009, p. 120).

Recent analysis of the media framing of geoengineering lends further support for the idea that transference idealization is a culturally sanctioned dynamic (Anshelm and Hansson 2014; Nerlich and Jaspal 2012); though some claim the varied and complex cultural framings involved are still emerging (Luokannen et al. 2014; Porter and Hulme 2013). Klein also offers a detailed analysis of how faith in geoengineering is created and sustained through a collusion of state sponsors, corporate backers and collective wishful thinking (2014, pp. 256–290; see also Erickson 2015).

Transference idealization is a way of disavowing individual and collective responsibility and the need for changes in everyday life and social order that threaten ontological security. It splits off feelings of guilt, anger, hope and responsibility, projecting them onto valorized others. For Dickinson and others, as well as facilitating inertia, transference idealization holds the political danger of passivity in the face of increasingly managerial or authoritarian governments in alliance with established corporate interests and technological infrastructure (Dickinson 2009; Klein 2014; Castree 2014).[2]

Self-Esteem and Consumerism

Increased self-esteem striving in the context of consumerism is Dickinson's second manifestation of distal defences. If we broadly accept the implications of a sociological concept of ontological security and an existential/social psychological perspective on terror management, we are 'beneficiaries of cheap fossil fuel' not just in a literal sense; but also in terms of how this energy form has underpinned the material infrastructure and the cultural worldview of 'advanced' consumer societies. In his introduction to *The Denial of Death* Becker notes that the 'ignoble heroics of whole societies… can be the viciously destructive heroics of Hitler's Germany or the plain debasing and silly heroics of the acquisition and display of consumer goods, the piling up of money and privileges that now characterizes whole ways of life' (1973, p. 7).

The quest for symbolic immortality is what motivates self-esteem striving according to Terror Management Theory, and as Dickinson attests, in Western society this 'could mean counterintuitive increases in status-driven consumerism, materialism, and other behaviors that increase carbon emissions' (2009, p. 37). Such an assertion rests on the assumption that we are heavily invested in the acquisition of consumer goods; and which sociological conceptions of 'consumerism' we accede to. On the one hand, consumerism is defined as 'a culturally manufactured, socially constructed and economically mandated desire for and preoccupation with endless acquisition of consumer goods and experiences' (Humphrey 2009, p. 42). This definition covers all the bases—social, cultural, economic, affective and cognitive. Although self-esteem is not explicitly labelled here, 'desire and preoccupation' suggests we are heavily invested personally. Widespread theories of consumer lifestyle suggest the individual is in thrall to the cultural logics of commodity fetishism, enmeshed in a hermetically sealed symbolic system of value and exchange. Campbell goes as far as to say that the activity of consuming:

> has become a kind of template or model for the way in which citizens of contemporary Western societies have come to view all their activities. Since … more and more areas of contemporary society have become assimilated to a 'consumer model' it is perhaps hardly surprising that the underlying metaphysics of consumerism has in the process become a kind of default philosophy for all modern life. (Campbell 2004, pp. 41–2)

There are plenty of examples of cultural logic that appear to support the idea that we adhere to the social, economic and cultural mandate of consumerism to the extent that it contributes to our emotional inoculation against ontological insecurity; or from a Terror Management Theory perspective, to hold mortality salience at bay: queues for the latest phones and tablets, followed by communal celebrations of purchases are a prominent example.

Bauman's acerbic analysis of the 'collateral damage' of consumerism offers an adjunct perspective, in highlighting the psychological and social implications of being 'locked out' of consumer lifestyles (2007a, b). Bauman's 'collateral casualties' of consumerism are the 'flawed consumers'—lacking resources that socially approved consumer activity requires (Bauman 2007b). They are 'people with no market value; they are the uncommoditized men and women, and their failure to obtain a status of proper commodity coincides with (indeed, stems from) their failure to engage in a fully-fledged consumer activity. They are *failed consumers*' (Bauman 2007b, pp. 31–2).

Bauman's analysis speaks of the idea that ontological security is deeply embedded in consumerism, but also wider recognition (symbolic immortality) in relation to employment, education, community life—those routinely or suddenly excluded point to that depth (e.g. Ehrenreich 2006). A search for meaning amidst the continuously shifting offer of commodities, provides in Bauman's analysis the 'sole acceptable – indeed badly needed and welcome – substitute' for collective identity and solidarity derived from world, supportive family and community' (2007a, b, p. 29).

While this line of argument works as a rousing polemic, it is clearly an exaggeration to raise consumerism to the status of sole narrative from which meaning can be derived. It is more reasonable to suggest that when anthropogenic ecological crisis further threatens the psychosocial basis for ontological security, we reach for whatever immortality-striving system we have available to us, and in increasingly manic ways. Simply stated, consumerism appears to provide a 'quick fix' to annul encroaching feelings of guilt, shame and anxiety and 'to deflect us from knowing about the underlying causes of our mounting anxiety' (Weintrobe 2010, p. 119).

Attempts to make theoretical and empirical links between Terror Management Theory and how we respond to knowledge of anthropogenic ecological degradation suggests that 'symbolic immortality' could at least potentially inhabit forms of conservation or activist behaviour (Dickinson 2009; Pienaar 2011; Vess and Arndt 2008). Some empirical findings have linked materialist sensibilities with a lack of concern for sustainability (Banarjee and McKeage 1994). Other experimental evidence links death-thought accessibility with increases in consumption-related behaviour (Kasser and Sheldon 2000; Arndt et al. 2004; Solomon et al. 2004). However, it would be simplistic to reduce any culture to a unipolar logic of consumerism and materialism (Rindfleisch and Burroughs 2004). There must be individual and group level variation in the extent to which self-esteem or ontological security is derived from materialist or consumerist frames. Terror Management Theory acknowledges that existential threat can trigger a wide range of responses, including more 'mature' defences akin to the psychoanalytic concept of sublimation; i.e. more positive and adaptive behaviour, though not necessarily tackling the source of anxiety directly.

Even within consumerism, the refraction of desire and longing involves complex impulses that cannot be reduced to a unidimensional measurement. Take the peculiar example of the film *Avatar* (2009, Dir. James Cameron). The film is set in the mid twenty-second century, and focuses on the relationship between a grey, drab, overdeveloped Earth and the newly discovered distant moon of Pandora and its inhabitants, the highly evolved Na'vi. It is a lush, green, biodiverse world—a thinly veiled earthly green paradise. A former marine travels to the planet as an avatar, initially to infiltrate the Na'vi and support the plans of a corporate-military complex to forcefully displace the Na'vi in order to extract the sought-after mineral 'unobtanium'. In the process, however, he falls in love with a Na'vi, and battles to defend Pandora. Avatar was one of the most expensively made films of all time and at the time of writing is *the* highest-grossing film, worldwide, of all time. This is a consumer experience millions have participated in, heavily advertised and merchandised, so it is in some ways a perfect exemplar of the 'underlying metaphysics' of consumerism; of the 'desire for and preoccupation with' mobilized by social structural, economic and cultural forces.

Nonetheless, the enormous success of the film and its ecological narrative has led some to suggest it signifies the dawning of a new consciousness (Brown 2011; Bernstein 2011). However, it can also be framed as potent example of splitting (Weintrobe 2010; Bo 2010). Jake Sully's travel between the two worlds is akin to our splitting of one messy, complicated earth into two separate ones—the dystopian Earth and the utopian Pandora. While in some ways the film displays environmental ethics writ large, this splitting is an escape from reality, a green phantasy, the unobtainability of which is then a source of further despondency and paralysis, the after effects of which are a paralyzing melancholy:

> While the film purports to be pro-environmental... the psychic message delivered by the story is about leaving the world. Our bodies and our planet are too broken. It is now about constructing virtual worlds through which to extend our fantasies of a place prior to the rift. Don the glasses and leave our world of plastic cups and sticky soda, and drift among the trees and exotic species likely to be endangered on our own planet. (Croker 2010, p. 42)

In other words, ecology as a consumable, Disneyfied phantasyland. The role of consumerism as a predominant narrative frame, one which appropriates sustainability, is developed further in the next chapter.

Outgroup Antagonism

Dickinson's third manifestation of distal defences is an increase in outgroup antagonism. Some time ago, Opotow and Weiss studied conceptions of justice involved in environmental conflicts, and specifically the utilization of moral exclusion to defend positions (Opotow and Weiss 2000). Moral exclusion 'rationalizes and justifies harm for those outside, viewing them as expendable, undeserving, exploitable, or irrelevant' (2000, p. 478). Although concerned with the stakeholder conflict over environmental issues such as air pollution, their analysis of the biased evaluation of groups usefully illustrates the distal defence of outgroup antagonism, and the related dynamics of disavowal, splitting and projection.

Alongside the exaggerated praising of ingroups, the authors claim that the biased evaluation of groups also involves the denigration of outgroups. This can involve condescension, derogation and dehumanization; projections that can engender the 'normalization and glorification of violence' towards outgroups, as well as expressions of a fear of contamination by 'them'. Such a fear is not surprising in the context of splitting and projection. The disavowed emotion or response is effectively disowned in the process of being projected onto others, so disgust and fear militate against its re-introjection on environmentalism replicate existing social distinctions.

Rustin, following Bion, similarly claims that anxieties induced by environmental crises may lead to defences patterned by a reversion to 'basic assumption groups' (2010, pp. 477–478). This phantasy-based articulation involves a disavowal of the capacity to deal with the source of anxiety reflectively and to take personal or collective responsibility. Patterns include the splitting of others into homogenously 'good' or 'bad' groups, allowing for the easy identification of enemies, upon whom problems are projected, 'and against whom antagonism can be mobilized' (2010, p. 477). Social psychological research, in which participants are prompted to consider some of the adverse consequences of climate change where they reside, lends support to this understanding (e.g. Fritsche et al. 2012; Devine-Wright et al. 2015). Such reminders increased levels of system justification and outgroup derogation, particularly among those who strongly identified with their nation. Others interpret similar activity more prosaically as 'blame-shifting': a technique of denial that allows moral disengagement by disavowing accountability and projecting it onto others, who are constructed as more responsible (Hamilton 2012; Norgaard 2011; Stoll-Kleeman et al. 2001).

Support for the conceptualization of outgroup antagonism as a distal defence also connects back to social psychological and sociological claims regarding more generalized forms of ontological security. In social psychology, it is generally understood that a threat to one's group identity can instigate an increase in the derogation of outgroups (Abram and Hogg 2006). The critical criminologist Jock Young associates collective shifts towards punitive response to crime and deviancy with 'widespread economic and ontological insecurity' (Young 2002, p. 263). The way he elaborates on this correlation elsewhere has close parallels with Rustin's idea of a reversion to 'basic assumption groups':

> Because of ontological insecurity there are repeated attempts to create a secure base. That is, to reassert one's values as moral absolutes, to declare other groups as lacking in value, to draw distinct lines of virtue and vice, to be rigid rather than flexible in one's judgements, to be punitive and excluding rather than permeable and assimilative. (Young, cited in Van Marle and Maruna 2010, p. 8)

Avowed environmentalists are not exempt from outgroup antagonism. The 'unconscious gains' derived from splitting and projection 'also manifests as those working in the field seek to claim and own the reparative aspects of the work, while disowning and projecting into others, particularly business organizations and the rich members of society, responsibility for socioecological degradation' (Mnguni, p. 122).

In Young's analysis, which overlaps significantly with meta-theorists of social change such as Beck, Bauman and Giddens, social institutions—such as the criminal justice system—take on a role in defending whole societies against the existential anxiety that arises when the social and cultural foundations of ontological security are threatened. Anthropogenic ecological degradation adds a further dimension of uncertainty, risk and threat; hence, the claims that it in effect it exacerbates social defences along similar lines.

Bolstered Worldview

The final distal defence in Dickinson's typology is the bolstering of existing worldviews. Although, as we have noted, distal defences are not necessarily destructive, they can be creative and, in some accounts, responding to the reality of death can be the basis for the creating of meaning, and therefore, a foundation for ontological security—at a personal, cultural and historical level. In the context of anthropogenic ecological crisis this means there is potential for bolstering environmentalist and ecocentric worldviews, particularly amongst those who already invest meaning or self-esteem. Bolstering a worldview is not the same as bolstering everyday practices—there is no straightforward elision between attitudes/values and behaviour. Nonetheless, in conjunction with the defences described above, the reinforcement of worldviews is likely to shape how one relates to anthropogenic ecological crisis as an issue that requires a significant response.

Worldviews are varied, but the beliefs and values that make them up also appear to assemble in reasonably distinct patterns. Of particular relevance to anthropogenic ecological crisis is research that suggests two contrasting worldviews: the dominant social paradigm and the new ecological paradigm (Dunlap et al. 2000). Each individual will, of course, adopt and adapt their own idiosyncratic assemblage, and negotiate this in relation to others in ways that are not necessarily consistent over time or in relation to what they do. But if a broad divergence does exist, there is a danger that the enactment of proximal defences will further entrench and polarize worldviews. Dickinson provides the following example,

> People who find self-esteem via materialism and an ideology of entitlement will probably buy more SUVs and become more antagonistic toward environmental causes and points of view, favoring suppression of the environmental movement and harsher penalties for the more radical protestors. In contrast, people who find self-esteem through humanist ideologies or environmentalism should become increasingly militant and vocal about their causes. (2009, p. 39)

The danger here is another layer of splitting—retaining righteousness for one's ingroup and expelling wrongdoing, for it to be absorbed by the 'others'. In terms of the specific activities of Dickinson's quote above, for example, the multiplicity of social identifications available means that there are many opportunities for 'green' or 'sustainable' consumption that facilitate consumerist and environmentalist versions of esteem; in a surplus of culture we invest in multiple, even contradictory, worldviews simultaneously.

A neoliberal consumerist worldview involves disavowing knowledge of how achieving the 'good life' for some is achieved at a cost for many others. Otherwise, the ontological security provided by the procession ιρarently endless supply of goods and services, and the narrative frames which collectively reassure, is imperiled. The communication of anthropogenic ecological crisis only amplifies the costs, and therefore, the amount of disavowal required. Denial at this level demands the splitting off of the desired commodity from its origins in labour, violence or degradation that entangles human and nonhuman others on a massive scale (Cudworth 2015).[3]

Cohen describes this state of affairs as a 'landscape where ends and means, landmarks and destinations, bear only an ironical relationship with each other' (Cohen 2001, p. 278). Ironic dissociation is a defence that enables the 'convenient fictions and evasions' of consumer capitalism to be perpetuated (Webb 2012, p. 119). In doing so, the commodity can take on a life as part of a meaningful constellation of objects, gestures, displays and activities—as documented by the many scholars of commodity fetishism following Marx (e.g. Carver 1975; Castree 2001). Only by obscuring, rationalizing and ironizing the practices and relations of production and consumption can a consumerist, economic growth oriented, meritocratic worldview—and a concomitant level of 'habitual forgetfulness'—be maintained (Billig 1999; Webb 2012).

Organizational Dynamics

To demonstrate claims that defence mechanisms are also intersubjective, dialogical and social processes, more empirical and theoretical work is needed; as Cohen noted, what were originally psychological concepts cannot simply be transposed to roles, relationships and politics (2001, p. 50). There is scope for a greater understanding of the psychosocial dynamics involved that can only come from closer observation and engagement. Norgaard's work is a good example of the kind of ethnographic/anthropological methodologies that have been called for to counter individualist psychological framing of research in this area (Capstick et al. 2014). What is lacking, however, even in Norgaard's work, is a sense of the specific intersubjective dynamics at work here. This is in part because the concepts of disavowal, splitting and projection are used in general and loose terms as carriers of cognitive or emotional content. *How* the process of organization works between and across people is yet to be examined in any detail.

Interestingly an earlier generation of psychoanalytic scholars utilized defence mechanism concepts to inform the analysis of 'socially organized defences' in organizations such as hospitals (e.g. Bion 1952; Jaques 1955, 1953; Menzies Lyth 1960). As yet, little of this generation's work informs socially-oriented accounts of defence mechanisms in the context of anthropogenic ecological crisis, but there is clearly some potential. Both Jacques and Menzies Lyth offered detailed descriptions of how collusive splitting and projection created shared external objects. They convey how a non-linear sequence or dialogue takes place, a 'two-way play' of 'projective and introjective identification' (Jaques 1953, p. 425).

The main issue here is not perhaps that this earlier generation needs to be reverently resurrected—rather that however complex and uncertain, the intersubjective and interpersonal dynamics ought to be at the heart of understandings of socially organized defence mechanisms; and, vitally, in considering how psychoanalytic observation might inform interventions—an issue which we now address in more detail.

Containing the Unbearable

> The contribution of psychoanalytic observers to these situations should be to track these unconscious structures of feeling (including their presence in their own minds) and to bring these sentient dimensions into public debate. They should not expect that their interventions will always be welcomed. (Rustin 2010, p. 478)

The task Rustin highlights here is akin to the therapeutic goal of psychoanalysis – to make unconscious dynamics conscious. No doubt it is a positive move to bring into awareness structures of feeling that encourage inertia, inaction and other responses that prevent or even escalate the social and psychological engagement with anthropogenic ecological crisis. Nonetheless, if this is the primary contribution of a psychoanalytic approach, there is still a need to detail the dynamics involved. At first glance Rustin is here advocating another form of the information defi-

critique of psychoanalysis

cit argument—where more knowledge about the problem is the key to resolving it. However, the adoption of a psychoanalytic perspective moves beyond the deficiencies of an information processing model. Again there is untapped potential in an earlier generation of psychoanalysts, particularly Wilfred Bion. His work on group experience is described above, but his account of the processes involved in thinking, and more specifically, projective identification, are also wholly relevant here.

Bion advanced Klein's work (1946) on projective identification in various directions, but it is perhaps his theory of 'containment' that hints at a more fully developed psychoanalytic contribution to public engagement with anthropogenic ecological degradation (Bion 1962, 1963).[4] For Bion, 'containment' is precisely the ability to convert unbearable states of mind into bearable ones. Although part of a broader account of mental functioning, it is a profoundly intersubjective concept, borne of clinical experience, that reflects Bion's reputation (along with Winnicott) as a 'true innovator of the *in between*' (Cartwright 2014, p. 2). Sayers summarizes the intersubjective nature of containment in the context of a therapeutic dyad:

> the 'realistic' projection of states of mind that are unbearable for the projector creates the possibility that they might still be borne by the mind of another, who receives them, converts them into a form that is more bearable ('detoxifies' them) and returns the more bearable form to the personality from which they originated (Sayers 1988, p. 140)

For Bion, following Klein, containment was understood in the context of projective identification in infancy, where 'good' and 'bad' elements of self and other (especially primary caregivers) are split, projected and introjected in early psychic life wrought with fantasy. The 'realistic' aspect of projective identification refers to the idea that the projection of unbearable states and re-introjection of more bearable ones is not just imagined by the projector, but is an actual intersubjective process: 'by no stretch of the imagination can this be understood as occurring in fantasy only' (Sandler 1988a, p. 19).

Bion's theory of containment, which has numerous parallels with Winnicott's account of the 'holding environment' (1964), speaks to the complexities of how the capacity to address 'unbearable' thoughts rely on conducive intersubjective dynamics. Understandably, containment has predominately been studied as it unfolds as a function of the analytic encounter (e.g. Bianchini et al. 2011; Billow 2000), or as part of early development in child-caregiver relationship (e.g. Douglas 2007; Fonagy et al. 2013). There has been some consideration of the wider implications of the concept of containment in relation to 'unbearable' psychological states beyond clinical settings, though still with a broadly therapeutic focus, noticeably in social work settings (e.g. Ruch 2007; Steckley 2012).

Mnguni offers a rare example of recent work that takes up this body of literature to make sense of an organization's attempt to become more sustainable, and how individuals use the dynamics of a social institution both creatively and defensively as anxiety arises (2010). In doing so, he appears to have provided some novel insights, such as the tendency for members of organizations and funding bodies to enter into 'collusive agreements' to enact and promote 'performances' of sustainability despite all involved parties being aware of their meaninglessness.[5]

If more of this kind of work can incorporate the conceptual detail we see in the earlier generation of 'organization psychoanalysis' there is the potential to more meaningfully flesh out how 'socially supportive' dynamics needed to hold and express anxiety might work. Ethnographies here might focus on people's perceptions of change, intersubjective and interpersonal dynamics, community and collective attempts to organize and respond creatively, the role of social support, shared narratives that develop or are challenged and so on. They might relatedly, as Mnguni suggests, help to develop action research and process facilitation methods that can inform interventions: 'Only then might people feel contained enough to be able to let go their defenses and bring to the task at hand their full capabilities, among them those for independent and critical thought and for mature relatedness' (2010, p. 133).

There is not the space to develop Bion's account further here, other than to note that his thinking stretches rudimentary uses of Freudian 'defences' to offer a glimmer of the possible value of psychoanalytic insight. If the implications of anthropogenic ecological degradation represent an 'unbearable' state of mind then they might depend on being suitably 'received' and 'converted' by others, rather than simply 'tracked'.

This is not to suggest that Bion's psychology of containment can easily be applied to larger group processes, but it might help understand the form of supportive social contexts Randall and others assert are the basis for meaningful intervention (Randall 2009, 2012). In terms of interventions, it also offers something more than Terror Management Theory's emphasis on 'better' information campaigning on environmental issues, and the need to manipulate mortality salience presentation to determine 'the most effective ways to structure climate change education' (Dickinson 2009, p. 39).

Creating Support Structures

The need to carry the social and cultural understanding of defence mechanisms, and the likelihood of resistance, over into implications for change, is perhaps best recognized by Randall's work. She summarizes her hopes for interventions inspired by social psychoanalytic understanding as follows:

> we need to create support structures that facilitate the process of mourning and provide containment for the anxieties that will inevitably be revealed. We need strategies that deal with the difficult issues of status and identity and a culture of stories and role-models that offers meaningful examples to identify with. (Randall 2009, p. 126)

The first point recognizes the need to do more than just expose and confront uncomfortable emotional responses in relation to anthropogenic ecological crisis but to provide a shared space for coming to terms with them. If anxiety is allowed to surface in a supportive social setting, it is more likely that it can be contained—channeled into more productive engagement instead of spurring on further defensive dynamics. Vitally, if we understand defence mechanisms as socially organized, at the personal, group and societal level, any hope for developing alternative responses must also be recognized and supported at those levels. If we follow the earlier tradition of group and organizational psychoanalysis, briefly described above, the importance of working at the level of recurring relationships is further reiterated. Randall's *Carbon Conversation* project is a practical attempt to develop and promote this kind of support structure at the group level, and positive results have been cautiously reported (Howell 2013; Randall 2012). Whether such projects can exceed their origins, the complexity and obdurateness of broader material and social structures is a critical question (Taylor Aiken 2015).

The second need identified by Randall in the quote above is closely related to the first but it deepens the role assigned to the relational and social dynamics involved in change. Status and identity are issues that have been raised here in relation to the maintenance of ontological security—particularly in stressing the importance of shared cultural frameworks as the basis for meaning and the ability to go on. Here, we also connected to Terror Management Theory, where responses to the universal experience of death-anxiety, which anthropogenic ecological crisis is taken to represent, are translated into culturally contingent attempts to maintain 'symbolic immortality'. Issues of status and identity loom large here, as do the 'culture of stories and role-models' that Randall identifies. As we have seen Terror Management Theory emphasizes the importance of role-models and worldviews that provide opportunities to create meaning that exceeds an individual's dread of finiteness.

In applying Terror Management Theory, Dickinson is cautious, but argues that 'if true change requires both heroic leadership and a cultural context for the heroism of many, a cultural world view that incorporates both innovation and idealization of the natural world is the logical immortality project and the best opportunity for heroism in these times' (Dickinson 2009, p. 11). Perhaps the contours of a cultural worldview as innovation/ idealization of the natural world are understandably opaque or prefigurative, but there are ongoing and newly emerging attempts to refigure the human-nonhuman relationship, and some of these transdisciplinary developments are the focus of subsequent chapters. The question, which cannot be answered with any certainty here, is 'whether the arguments that need to be made for 'belonging' and 'relationship', in all their various forms, and for the social systems and practices necessary to sustain those, can gain sufficient traction to make a difference' (Rustin 2014, pp. 157–8).

There is still plenty of scope for research and interventions inspired by psychoanalysis to help us make sense of how as human beings we engage with 'knowing and not knowing' about anthropogenic ecological crisis. Following a systematic review of literature addressing public perceptions of climate change, Capstick et al. point to the problems of 'individualistic research framings' of 'the public', which individualize collective problems and close down the potential for dialogue between citizens (2014, p. 21). They conclude by calling for the greater deployment of qualitative work, including anthropological/ethnographic and qualitative longitudinal methodologies to elucidate a 'depth of insight' that goes beyond stated attitudes, values and opinions. Similar calls are made for work in this area to examine the social and relational contexts in which meanings are negotiated, or to paraphrase Jaspal et al. the discursive and sense-making practices of non-elite actors (Jaspal et al. 2012). Though it could be added, following discussions in this book, that research exploring the relational and discursive dynamics of incumbent elite actors is equally vital (Geels 2014). Relatively few studies have inquired into how popular discourses and narratives related to climate change are negotiated socially.

Yet we must be wary of the subsequent contention 'that a paucity of such research has curtailed the empirical basis for *fashioning messages* that could sidestep divisive, didactic, and ideologically driven climate change discourses' (Hanson-Easey et al. 2015, p. 220; emphasis added). To assume that the significance of work in this area should largely be thought of in terms of 'refashioning messages' is to miss the extent to which it raises our understanding of the social, or more accurately the psychosocial, to a different level.

Reflections on Knowing and Not Knowing

Knowing and not knowing about anthropogenic ecological crisis has been presented here as a dynamic process involving the psychosocial organization of defence mechanisms: as 'collective phenomena, social defenses come about as members collude to use organizational processes to reinforce their individual defense mechanisms' (Mnguni 2010, p. 123). Moving forward, it is imperative that we do not incorporate proximal and distal defences into a growing list of 'psychological barriers'. The shared, social collective dimension of defences of disavowal, splitting and projection means they are simultaneously psychological, intersubjective and social, and it makes no sense to talk about defences as 'internal' rather than 'external'. This is a distinction many well-respected commentators make, but it reifies the idea that defence mechanisms are another psychological barrier, rather than an ongoing psychosocial accomplishment.

If one is willing to make the leap of faith, a psychoanalytic perspective is important because it suggests that when certain states or experiences are disavowed, they do not simply 'disappear without a trace' (Layton 2014, p. 166). They are repressed, projected onto others but reappear in related forms: 'The split-off states, however, continue all the while to push for expression, not only in physical symptoms… but in characterological adaptations… and in particular kinds of relational scenarios' (Layton 2014, p. 166). This makes a psychodynamic approach different from a discursive one, where falling in with normative discourses or rebelling against them do not appear to have any tangible psychic consequences; or a straightforward sociological one, where the only discernable 'cost' of defence mechanisms in the context of anthropogenic ecological crisis is that an effective societal response is held back.

Thinkers inspired by psychoanalysis as varied as Adorno, Lacan, Lasch, Butler, Gramsci, Althusser and Laplanche have long made overlapping claims about the relational and social dynamics of unconscious life, and the extent to which our identity is split, enigmatic, and subjugated, preemptive of any consciousness of these ruptures and fissures (Craib 1998). What they share is the sense that cultural malaise can be translated into individual disaffection in ways we cannot fully understand or articulate; 'that we can suffer from many problematic aspects of the social order without realizing the actual source of discordant experience' (Hollander 2009, p. 3). It is just possible that comprehending anthropogenic ecological degradation—an existential crisis of planetary proportions—requires just such a theoretical framework.

The shift between proximal and distal defences is also a potentially vital one. According to Terror Management Theory, as anthropogenic ecological crisis encroaches further on everyday life, it will generate more opportunities for mortality salience and ontological insecurity. The threat this poses increases the call for distal defences, and with it increasingly manic attempts to maintain ontological security and symbolic immortality. Social and cultural vehicles for these defences are abundant, providing ample opportunities for their further validation. These actions lessen the chance of meaningful engagement with the issues in question. Hollander captures this apparent paradox nicely in general terms:

Paradoxically, the more we defend ourselves against fears of being overwhelmed by reverting to denial, splitting, projection and disavowal, the less our chances of adaptively mobilizing and of experiencing our own sense of agency in the struggle to create a more harmonious relationship between humanity and the rest of nature. Not knowing allies with destructiveness, and we wind up being part of a bystander population that ultimately reinforces the real threat to survivability. (Hollander 2009, p. 7)

The attempt across numerous disciplines to look both 'deeper', by considering less rational motivations below the surface, and 'wider', by considering the social dynamics contributing to affective responses to knowing about anthropogenic ecological degradation, is a welcome move. In the next two chapters, we pursue these dynamics further, via an exploration of narrative approaches.

Notes

1. Environmentalist narratives can also disavow the dependency of the relationship between humans and the rest of nature in the popular conception of humans as some kind of parent of the earth – its (potential) savior or healer, with attendant images of the earth in our hands. Dependency and vulnerability in our relationship to the rest of nature is sequestered even from this narrative, denying the fact that 'it is we humans who are fragile and vulnerable and the earth that is hearty and powerful... an earth that, if pushed too far, has ample power to rock, burn, and shake us off completely' (Klein 2014, p. 285).
2. Earlier thinkers like Andre Gorz offered similar warnings: 'The rejection of technofascism does not arise from a scientific understanding of the balances of nature, but from a cultural and political choice. Environmentalists use ecology as the lever to push forward a radical critique of our civilization and our society. But ecological arguments can also be used to justify the application of biological engineering to human systems' (Gorz 1979, p. 17).
3. Though moves towards demystifying the material and social networks involved in producing commodities, such as Fairtrade, threaten to 'remove the veil' of consumer capitalism (Hudson and Hudson 2003), it remains adept at absorbing and appropriating such movements into that worldview.
4. Thanks to Wendy Hollway for making this initial connection to Bion's work, in reviewing an earlier draft of this chapter.
5. Perversely the more that the consequences of the Anthropocene are experienced directly, and mediated as such, the more opportunities there are for

this kind of research. The ongoing California drought, for example, in its fourth consecutive year at time of writing, is now publically presented as an outcome of climate change by senior figures including the state's governor Jerry Brown (McCarthy 2015); and changes in behaviour related to water use have been encouraged and demanded (e.g. Fraumeni 2015).

References

Abrams, D., & Hogg, M. A. (2006). *Social identifications: A social psychology of intergroup relations and group processes*. London: Routledge.

Anshelm, J., & Hansson, A. (2014). The last chance to save the planet? An analysis of the geoengineering advocacy discourse in the public debate. *Environmental Humanities, 5*, 101–123.

Arndt, J., Solomon, S., Kasser, T., & Sheldon, K. M. (2004). The urge to splurge: A terror management account of materialism and consumer behavior. *Journal of Consumer Psychology, 14*(3), 198–212.

Banerjee, B., & McKeage, K. (1994). How green is my value: Exploring the relationship between environmentalism and materialism. In C. T. Allen & D. Roedder (Eds.), *Advances in consumer research* (Vol. 21, pp. 147–152). Association for Consumer Research: Provo.

Bauman, Z. (2007a). *Consuming life*. London: Wiley.

Bauman, Z. (2007b). Collateral casualties of consumerism. *Journal of Consumer Culture, 7*(1), 25–56.

Becker, E. (1973). *The denial of death*. New York: Simon and Schuster.

Bernstein, J. S. (2011). On the edge: Borderland consciousness and Avatar. An emergent myth of our time. *Quadrant, 41*(1), 5–9.

Bianchini, B., Dallanegra, L., & O'flaherty, R. (2011). Reflections on the container-contained model in couple psychoanalytic psychotherapy. *Couple and Family Psychoanalysis, 1*(1), 69–80.

Billig, M. (1999). Commodity fetishism and repression: Reflections on Marx, Freud and the psychology of consumer capitalism. *Theory & Psychology, 9*(3), 313–329.

Billow, R. M. (2000). Relational levels of the "container-contained" in group therapy. *Group, 24*(4), 243–259.

Bion, W. R. (1952). Group dynamics: A re-view. *International Journal of Psychoanalysis, 33*, 235–247.

Bion, W. R. (1962). *Learning from experience*. London: Heinemann.

Bion, W. R. (1963). *Elements of psychoanalysis*. London: Heinemann.

Blake, J. (1999). Overcoming the 'value-action gap' in environmental policy: Tensions between national policy and local experience. *Local Environment,* 4(3), 257–278.

Bo, Y. (2010). Who is in charge of the body and mind of Jake's Avatar? A cultural and cognitive approach to identity construction. *Contemporary Foreign Languages Studies, 11,* 010.

Boykoff, M. T., & Goodman, M. (2009). Conspicuous redemption: Promises and perils of celebrity involvement in climate change. *Geoforum, 2009*(40), 395–406.

Brown, G. (2011). Avatar, the movie: Awakening between two worlds. *Quadrant, 41*(1), 1–4.

Campbell, C. (2004). I shop therefore I know that I am: The metaphysical basis of modern consumerism. In K. Ekstrom & H. Brembeck (Eds.), *Elusive consumption: Tracking new research perspectives* (pp. 27–44). Oxford: Berg.

Capstick, S., Lorenzoni, I., Corner, A., & Whitmarsh, L. (2014). Prospects for radical emissions reduction through behavior and lifestyle change. *Carbon Management, 5*(4), 429–445.

Cartwright, D. (2014). *Containing states of mind: Exploring Bion's 'container model' in psychoanalytic psychotherapy.* London: Routledge.

Carver, T. (1975). Marx's commodity fetishism. *Inquiry, 18*(1), 39–63.

Castree, N. (2001). Commodity fetishism, geographical imaginations and imaginative geographies. *Environment and Planning A: International Journal of Urban and Regional Research, 33*(9), 1519–1525.

Castree, N. (2014, November 14). Dangerous knowledge and global environmental change: Whose epistemologies count? EnviroSociety. www.envirosociety.org/2014/11/dangerous-knowledge-and-global-environmental-change. Accessed 19 Dec 2015.

Cohen, S. (2001). *States of denial: Knowing about atrocities and suffering* New York: Wiley.

Cohen, S. (2013). Discussion: Climate change in a perverse culture. In S. Weintrobe (Ed.), *Engaging with climate change: Psychoanalytic and interdisciplinary perspectives* (pp. 72–79). London: Routledge.

Craib, I. (1998). *Experiencing identity.* London: Sage.

Cudworth, E. (2015b). Killing animals: Sociology, species relations and institutionalized violence. *The Sociological Review, 63*(1), 1–18.

Devine-Wright, P., Price, J., & Leviston, Z. (2015). My country or my planet? Exploring the influence of multiple place attachments and ideological beliefs upon climate change attitudes and opinions. *Global Environmental Change, 30,* 68–79.

Dickinson, J. L. (2009). The people paradox: Self-esteem striving, immortality ideologies, and human response to climate change. *Ecology and Society, 14*(1), 34.

Douglas, H. (2007). *Containment and reciprocity: Integrating psychoanalytic theory and child development research for work with children.* London: Routledge.

Dunlap, R., Liere, K. V., Mertig, A., & Jones, R. E. (2000). Measuring endorsement of the new ecological paradigm: A revised NEP scale. *Journal of Social Issues, 56*(3), 425–442.

Ehrenreich, B. (2006). *Bait and switch: The futile pursuit of the corporate dream.* London: Granta.

Erickson, M. (2015). *Science, culture and society: Understanding science in the 21st century* (2 ed.). Cambridge: Polity.

Firestone, R. W. (1994). Psychological defences against death anxiety. In R. A. Neimeyer (Ed.), *Death anxiety handbook: Research, instrumentation, and application* (pp. 214–241). Washington, DC: Taylor and Francis.

Fivush, R. (2010). Speaking silence: The social construction of silence in autobiographical and cultural narratives. *Memory, 18*(2), 88–98.

Fonagy, P., Steele, M., Steele, H., Leigh, T., & Kennedy, R. (2013). The predictive specificity of the adult attachment interview and pathological emotional development. In S. Goldberg, R. Muir, & J. Kerr (Eds.), *Attachment theory: Social, developmental, and clinical perspectives* (pp. 233–278). New Jersey: The Analytic Press.

Fraumeni, P. (2015, April 21). Can human behaviour fix the California drought? University of Toronto. http://www.research.utoronto.ca/can-human-behaviour-fix-the-california-drought/. Accessed 18 Dec 2015.

Fritsche, I., Cohrs, J. C., Kessler, T., & Bauer, J. (2012). Global warming is breeding social conflict: The subtle impact of climate change threat on authoritarian tendencies. *Journal of Environmental Psychology, 32*, 1–10.

Gifford, R. (2011). The dragons of inaction: Psychological barriers that limit climate change mitigation and adaptation. *American Psychologist, 66*(4), 290–302.

Gorz, A. (1979). *Ecology as politics.* London: Southend Press.

Hamilton, C. (2012). What history can teach us about climate change denial. In S. Weintrobe (Ed.), *Engaging with climate change: Psychoanalytic & interdisciplinary perspectives* (pp. 16–32). London: Routledge.

Hanson-Easey, S., Williams, S., Hansen, A., Fogarty, K., & Bi, P. (2015). Speaking of climate change: A discursive analysis of lay understandings. *Science Communication, 37*(2), 217–239.

Hollander, N. C. (2009). When not knowing allies with destructiveness: Global warning and psychoanalytic ethical non-neutrality. *International Journal of Applied Psychoanalytic Studies, 6*(1), 1–11.

Howell, R. A. (2013). It's not (just) "the environment, stupid!" Values, motivations, and routes to engagement of people adopting lower-carbon lifestyles. *Global Environmental Change, 23*(1), 281–290.

Hudson, I., & Hudson, M. (2003). Removing the veil? Commodity fetishism, fair trade, and the environment. *Organization & Environment, 16*(4), 413–430.

Humphery, K. (2009). *Excess: Anti-consumerism in the West*. Cambridge: Polity.

Jaques, E. (1953). On the dynamics of social structure: A contribution to the psychoanalytic study of social phenomena deriving from the views of Melanie Klein. *Human Relations, 6*, 3–24.

Jaques, E. (1955). Social systems as a defence against persecutory and depressive anxiety. In M. Klein, P. Heimann, & R. E. Money-Kyrle (Eds.), *New directions in psychoanalysis* (pp. 478–498). London: Tavistock Publications.

Jaspal, R., Nerlich, B., & Koteyko, N. (2012). Contesting science by appealing to its norms: Readers discuss climate science in The Daily Mail. *Science Communication, 25*, 383–410.

Jordan, M. (2009). Nature and Self—An Ambivalent Attachment? *Ecopsychology, 1*(1), 26–31.

Kasser, T., & Sheldon, K. M. (2000). Of wealth and death: Materialism, mortality salience, and consumption behavior. *Psychological Science, 11*, 348–351.

Keene, J. (2013). Unconscious obstacles to caring for the planet: Facing up to human nature. In S. Weintrobe (Ed.), *Engaging with climate change: Psychoanalytic and interdisciplinary perspectives* (pp. 144–159). London: Routledge.

Klein, M. (1946). Notes on some schizoid mechanisms. *International Journal of Psychoanalysis, 27*, 99–110.

Klein, N. (2014). *This changes everything*. Harmondsworth: Penguin.

Langford, I. H. (2002). An existential approach to risk perception. *Risk Analysis, 22*, 101–120.

Layton, L. (2014). Some psychic effects of neoliberalism: Narcissism, disavowal, perversion. *Psychoanalysis, Culture and Society, 19*, 161–178.

Lertzman, R. (2015). *Environmental melancholia: Psychoanalytic dimensions of engagement*. London: Routledge.

Lorenzoni, I., Nicholson-Cole, S., & Whitmarsh, L. (2007). Barriers perceived to engaging with climate change among the UK public and their policy implications. *Global Environmental Change, 17*(3), 445–459.

Luokkanen, M., Huttunen, S., & Hildén, M. (2014). Geoengineering, news media and metaphors: Framing the controversial. *Public Understanding of Science, 23*(8), 966–981.

McCarthy, T. (2015, April 5). California governor tells climate change deniers to wake up. The guardian. https://www.theguardian.com/us-news/2015/apr/05/california-governor-drought-climate-change-dianne-feinstein. Accessed 12 Dec 2015.

McKibben, B. (2012, August 2). Global warming's terrifying new math. *Rolling Stone,* Issue 1162. http://www.rollingstone.com/politics/news/global-warmings-terrifying-new-math-20120719#ixzz3SrbRS2SV. Accessed 18 Dec 2015.

Menzies Lyth, I. E. P. (1960). The functions of social systems as a defence against anxiety: A report on a study of the nursing service of a general hospital. *Human Relations, 13*(2), 95–121.

Mishan, J. (1996). Psychoanalysis and environmentalism: First thoughts. *Psychoanalytic Psychotherapy, 10*, 59–70.

Mnguni, P. P. (2010). Anxiety and defense in sustainability. *Psychoanalysis, Culture and Society, 15*(2), 117–135.

Nerlich, B., & Jaspal, R. (2012). Metaphors we die by? Geoengineering, metaphors, and the argument from catastrophe. *Metaphor and Symbol, 27*(2), 131–147.

Norgaard, K. M. (2011). *Living in denial: Climate change, emotions and everyday life*. Cambridge: MIT Press.

Opotow, S., & Weiss, L. (2000). New ways of thinking about environmentalism: Denial and the process of moral exclusion in environmental conflict. *Journal of Social Issues, 56*, 475–490.

Pienaar, M. (2011). An eco-existential understanding of time and psychological defenses: Threats to the environment and implications for psychotherapy. *Ecopsychology, 3*(1), 25–39.

Porter, K. E., & Hulme, M. (2013). The emergence of the geoengineering debate in the UK print media: A frame analysis. *The Geographical Journal, 179*(4), 342–355.

Pyszczynski, T., Greenberg, J., & Solomon, S. (1999). A dual-process model of defense against conscious and unconscious death-related thoughts: An extension of terror management theory. *Psychological Review, 106*, 835–845.

Randall, R. (2005). A new climate for psychotherapy? *Psychotherapy and Politics International, 3*, 165–179.

Randall, R. (2009). Loss and climate change: The cost of parallel narratives. *Ecopsychology, 3*, 118–129.

Randall, R. (2012). Fragile identities and consumption: The use of 'Carbon Conversations' in changing people's relationship to 'stuff'. In M.-J. Rust & R. Totton (Eds.), *Vital signs: Psychological responses to ecological crisis* (pp. 225–238). London: Karnac Books.

Rindfleisch, A., & Burroughs, J. E. (2004). Terrifying thoughts, terrible materialism? Contemplations on a terror management account of materialism and consumer behavior. *Journal of Consumer Psychology, 14*(3), 219–224.

Ruch, G. (2007). Reflective practice in contemporary child-care social work: The role of containment. *British Journal of Social Work, 37*(4), 659–680.

Rustin, M. (2010). Looking for the unexpected: Psychoanalytic understanding and politics. *British Journal of Psychotherapy, 26*(4), 472–479.

Rustin, M. (2012). How is climate change an issue for psychoanalysis? In S. Weintrobe (Ed.), *Engaging with climate change: Psychoanalytic and interdisciplinary perpsectives* (pp. 170–185). London: Routledge.

Rustin, M. (2014). Belonging to oneself alone: The spirit of neoliberalism. *Psychoanalysis, Culture & Society, 19*(2), 145–160.

Sandler, J. (1988a). The concept of projective identification. In J. Sandler (Ed.), *Projection. Identification, projective identification* (pp. 13–26). London: Karnac Books.

Sandler, J. (1988b). Foreword. In J. Sandler (Ed.), *Projection. Identification, projective identification* (pp. 1–3). London: Karnac Books.

Sayers, S. (1998). *Marxism and human nature.* Hove: Psychology Press.

Searles, H. (1960). *The nonhuman environment in normal development and in schizophrenia.* New York: International Universities Press.

Searles, H. (1972). Unconscious processes in relation to the environmental crisis. *Psychoanalytic Review, 59*, 361–374.

Solomon, S., Greenberg, J., & Pyszczynski, T. A. (2004). Lethal consumption: Death-denying materialism. In T. Kasser & A. D. Kanner (Eds.), *Psychology and consumer culture: The struggle for a good life in a materialistic world* (pp. 127–146). Washington, DC: American Psychological Association.

Steckley, L. (2012). Touch, physical restraint and therapeutic containment in residential child care. *British Journal of Social Work, 42*(3), 537–555.

Stoll-Kleemann, S., O'Riordan, T., & Jaeger, C. C. (2001). The psychology of denial concerning climate mitigation measures: Evidence from Swiss focus groups. *Global Environmental Change, 11*(2), 107–117.

Taylor Aiken, G. (2015). (Local-) community for global challenges: Carbon conversations, transition towns and governmental elisions. *Local Environment, 20*(7), 764–781.

Van Marle, F., & Maruna, S. (2010). 'Ontological insecurity' and 'terror management': Linking two free-floating anxieties. *Punishment and Society, 12*(1), 7–26.

Vess, M., & Arndt, J. (2008). The nature of death and the death of nature: The impact of mortality salience on environmental concern. *Journal of Research in Personality, 42*, 1376–1380.

Webb, J. (2012). Climate change and society: The chimera of behaviour change technologies. *Sociology, 46*(1), 109–125.

Weintrobe, S. (2010b). Engaging with climate change means engaging with our human nature. *Ecopsychology, 2*(2), 119–120.

Winnicott, D. W. (1964). *The child, the family, and the outside world.* Harmondsworth: Penguin.

Yalom, I. D. (1980). *Existential psychotherapy.* New York: Basic Books.

Young, J. (2002). Critical criminology in the twenty-first century: Critique, irony and the always unfinished. In K. Carrington & R. Hogg (Eds.), *Critical criminology: Issues, debates, challenges* (pp. 251–274). London: Taylor and Francis.

9

'Its All Folded into Normalcy': Narratives and Inaction

Introduction

…What happens is of little significance compared with the stories we tell ourselves about what happens. Events matter little, only stories of events affect us. (Rabih Alameddine, *The Hakawati*)

To hell with facts! We need stories! (Ken Kesey)

In 2009, following a successful crowdfunding appeal, Paul Kingsnorth and Dougland Hine published a small pamphlet called the *Uncivilization Manifesto*. It was the authors' initial attempt to articulate their feeling 'that contemporary literature and art were failing to respond honestly or adequately to the scale of our entwined ecological, economic and social crises' (The Dark Mountain Project 2016). It struck a chord. Responses to the manifesto, and conversations it inspired, soon grew into The Dark Mountain Project (DMP). The project's website describes it as

© The Author(s) 2016   **175**
M. Adams, *Ecological Crisis, Sustainability and the Psychosocial Subject*,
DOI 10.1057/978-1-137-35160-9_9

a network of writers, artists and thinkers who have stopped believing the stories our civilisation tells itself. We see that the world is entering an age of ecological collapse, material contraction and social and political unravelling, and we want our cultural responses to reflect this reality rather than denying it. (The Dark Mountain Project 2016)

The DMP has since held numerous public events, festivals, and published annual volumes of poetry, art, short fiction and non-fiction essays. It has attracted a substantial amount of both support and criticism from environmentalists, journalists and social commentators (e.g., Gray 2009; Monbiot 2010; Smith 2014a; Townsend 2010); as well as being the subject of a handful of academic articles (e.g., Hoggett 2011); and at least one PhD thesis to date (Graugaard 2014).

What I found striking about the *Uncivilization Manifesto* back in 2009, and subsequent volumes of art and writing (e.g., Kingsnorth and Hine 2010a; Hunt, Kingsnorth & Wheeler 2014; Du Cann et al. 2015) was the explicit emphasis on the importance of narrative and storytelling in making sense of how and why people respond to knowing about anthropogenic ecological degradation in the ways we do, individually and collectively. This was a relatively neglected issue in the psychological and sociological analyses of climate change communication and sustainability at that time, a literature I was exploring (and still largely overlooked today, according to George Marshall 2014b).

The authors explicitly dissociated themselves from both apocalyptic and technological-fix narratives in relation to anthropogenic ecological degradation, because, they argued, both circumscribe practical and imaginative responses that might genuinely unsettle dominant narratives. Both narratives validate the deceptive claim 'that life without the components of our current way of living is simply unlivable. That the future will give us either unbroken progress or apocalypse, and there are no spaces between' (Kingsnorth and Hine 2010b, p. 3). Instead, the authors claimed a space for dread, grief and loss, not as the basis for 'saving the world' or ordered progressive change, but as a more ambiguous occupation of 'the foothills of some dark and uncharted range' (Kingsnorth and Hine 2010b, p. 3). 'Uncivilization' is not a place, goal, ideal or political position, but a process 'of unlearning the assumptions, the founding narratives, of our civilization' (Kingsnorth and Hine 2011, p. 3).

Contributors consistently reiterate the Project's overriding preoccupation—the narrative framing of ecological degradation and the human-nonhuman nature relationship, grounded in a shared understanding of the fundamental importance of narrative for meaningful human life. The first four (of eight) principles of the *Uncivilization Manifesto* are especially pertinent to the issue of narrative:

1. We live in a time of social, economic and ecological unravelling. All around us are signs that our whole way of living is already passing into history. We will face this reality honestly and learn how to live with it.
2. We reject the faith which holds that the converging crises of our times can be reduced to a set of 'problems' in need of technological or political 'solutions'.
3. We believe that the roots of these crises lie in the stories we have been telling ourselves. We intend to challenge the stories which underpin our civilization: the myth of progress, the myth of human centrality, and the myth of our separation from 'nature'. These myths are more dangerous for the fact that we have forgotten they are myths.
4. We will reassert the role of storytelling as more than mere entertainment. It is through stories that we weave reality.

The Dark Mountain Project's emphasis on narrative resonates with a growing body of literature: work across a range of disciplines that declares an interest in the power of storytelling and narrative frames in conveying the realities of ecological degradation and a warming climate, not least as a tool for mobilizing people to act. Institutions large and small have developed interventions, published reports and staged events that make narrative the focus of climate change communication in particular.[1]

This chapter attempts to explore social science interpretations of the role of narrative in relation to the ecological crisis, and broader processes of social and subjective change. More specifically, it is an exploration of the role of narrative in conveying the reality of the ecological crisis; and, inseparably, the role of narrative in orientating collective and individual responses, actually and potentially. In other words, in shaping how we come to know about the ecological crisis as a crisis (or not) as well as how (and if) we should and could respond to this knowledge.

Emphasizing the influence of narrative here is not intended to suggest that there is an obvious or unidirectional causal connection between narratives, whatever we take that word to mean, and human experience and action. As we have seen in previous chapters, the psychosocial dynamics involved in knowing (and not knowing) about ecological crisis, and the relationship between everyday life, subjectivity and social dynamics is far more protean. While a focus on narrative might momentarily isolate and illuminate its constitutive role, our discussion will try to attend to the ways in which narrative is undoubtedly embedded and embodied in the contours of everyday life as so far described here; and, vitally, the ways in which narrative constructions engage with non-narrative reality.

We Need Stories

It is in telling our stories that we give ourselves an identity. (Ricoeur 1985, p. 214)

A prevalent and powerful idea to emerge from psychology and the social sciences in the last fifty years is that narratives are essential ingredients of socially meaningful subjectivity (e.g., Ricoeur 1984; Fisher 1987; Bruner 1990; Sarbin 1986). Some go as far as to identify a 'narrative turn' in the social sciences and humanities (Czarniawska 2004), as the importance acceded to narrative in the construction of meaning became evident across a range of disciplines including anthropology, education, psychotherapy, social psychology, sociology and human geography (e.g., Atkinson 1997; Berger and Quinney 2005; Brown et al. 1996; Gergen and Gergen 1986; Josselson and Lieblich 1999).

Narratives, stated at their simplest, are written or spoken accounts of connected events; as such the word narrative is interchangeable with the word story. Beyond basic definitions, there is no consensus on the meaning of the term 'narrative' (Chandler et al. 2004), so there is little chance of agreement on the basic tenets of a narrative turn – but I will attempt to provide a few pointers nonetheless. First, narratives are psychologically (biologically for some) fundamental, in that, human beings use a narrative to make sense of experience, in ways that incorporate intuitive, pre-conscious, interpersonal, group and societal sense-making (Gergen 2005; Lakoff 2010).[2]

Lakoff's position is that *all* knowledge, emotion, imagery, language, thinking and talking depends upon a limited number of 'frames' to be experienced and communicated as meaningfully coherent. Furthermore, the repeated utilization of associative connections between language and emotion within a frame literally embodies the reality of that particular frame, building up neural activity and laying down frames in 'brain circuitry'. Frames are therefore inevitable, as is their determination of our perception of reality, and hard to change once established in associative narratives. We therefore formulate and employ stories as part of the process of comprehending and organizing our experience of the world. The idea that we have a universal psychological tendency to narrate stems from various claims: about how humans experience time (Ricoeur 1984); about the dynamics of early child development (McAdams 2005; Mandler 1984); of the nature of language-use and social life (Bruner 1987); and apparent correlations with neurochemical processes that suggest a biological proclivity underpinning perception and meaning-making in humans (Lakoff 2010).

Second, narratives are considered important as internal and interpersonal dynamics that *construct* or co-construct meaning—they help explain and evaluate action within a shared system of symbolization, interweaving internal and external life into a coherent whole that we invest ourselves in. Narratives are not the icing on the cake of already established physiological and psychological capacities, they are fundamentally involved in *organizing* perception, representation, and interpretation of events, in specifically ordered ways (Lakoff 2010). It is in this sense that narratives are a vital conduit for individual expression and engagement with others—the basis for a meaningful agency. It is also why coherent identity is understood *as* narrative identity, defined by Widdershoven as 'the unity of a person's life as it is experienced and articulated in stories that express this experience' (1993, p. 7).

Third, and linking back to an important concept developed in previous chapters, narratives are also the primary vehicles through which we maintain a sense of ontological security. The ability to go on is here intimately linked with the temporal sequencing that narrative permits—

to 'imbue a sequence of actions with causal links that explain why one action follows another' (Fivush 2010, p. 89). The larger ontological project is the construction of a narrative identity; a formulation 'that selectively reconstructs the past and imagines the future as an integrated temporal whole, to provide life with meaning and purpose and situate the person's imagined life trajectory within a recognizable societal niche' (McAdams 2005, p. 243). A narrative structure warrants the persistence of a self as an entity over time. The link between a 'workable sense of personal persistence' and profound ontological security is perhaps nowhere clearer than in the remarkable series of studies by Chandler and colleagues, and the key finding that 'failures to warrant self-continuity are strongly associated with increased suicide risk in adolescence' (Chandler et al. 2003).[3]

This leads to a fourth general point regarding the importance attributed to a narrative. Taken generally, the narrative turn does not suggest that we simply create narratives as we go, like a train laying down tracks immediately in front of it as it hurtles forward on open terrain. Many accounts of narrative fundamentally acknowledge the role of the social and cultural in shaping the ways we narratively make sense of the world and our place in it (Gergen and Gergen 1988; Gergen 1991). Although asserted as a space for creativity and autonomy, agency yoked to narrative is not, then, unfettered and unbounded. There is an understanding that the will to meaningfully story the world arises in a world that is already storied. To derive ontological security from the narration of one's existence, it must chime with others' attempts to do so—it must be articulable 'according to the society's implicit understandings of what counts as a tellable story, a tellable life' (McAdams 2005, p. 243). To be 'tellable', a narrative must be appropriately recognised and legitimized by others; and, vitally, capable of being materially resourced and reinforced as and when necessary. The alternative, stated bluntly, is to *not* make sense to self and others – a state of ontological insecurity.

Once established as a fundamental unit of analysis, narrative scholars and researchers in the social sciences and humanities emphasize how some narratives come to dominate over others, how change occurs, and, of course, the identification of prevalent narratives socially, culturally and historically (McAdams 1993). Analysis of power and inequality is involved here, not least in the sense that while some stories are privileged, others are excluded, marginalized, denigrated or mocked; and in the fact that some members of a society have readier access to 'canonical' narratives than others (Fivush 2010). A longstanding feminist argument, for example, is that many women in Western societies 'have been deprived of the narratives, or the texts, plots, or examples, by which they might assume power over—take control over—their lives' (Heilbrun, cited in McAdams 2005, p. 250). However, it is clearly necessary to avoid a simplistic understanding of the relationship between individuals and culture (Gjerde 2004). In accepting that the intersecting dynamics of agency, power, narrative and culture are complex, the 'resultant compromise', for McAdams and others, is to view narrative identity as a psychosocial construction: 'Self and culture come to terms with each other *through* narrative' (McAdams 2005, p. 252; emphasis added).

At the most general level, the narrative turn acknowledges choice and agency, even if they are largely understood as the appropriation of stories limited by socially and culturally prescribed possibilities. This might not sound much like 'genuine' agency. However, even within the parameters of the narrative life stages of Western culture, changes that emerge from, and at least potentially impact significantly on, people's lived experience are discernable. Fivush considers the example of the ages at which childbirth is considered 'normal' as an example of the cultural variability of narrative framing, even in an area that at first glance may appear to be biologically mandated: 'Although life stages may be heavily based on biological factors (e.g., puberty, childbearing), cultures modulate these biological considerations in forming social expectations. For example, with the advent of the second wave of the women's movement, the age of childbirth is now a much wider culturally acceptable window than previously' (2010, p. 93). While many aspects of gender roles may still be firmly entrenched, and changes elsewhere patchy and piecemeal (McNay 1999), or dependent on intersectional inequalities (Yuval-Davis 2006), the change highlighted here is a result of a complex mix of active collective and individual resistance, subtle changes in social attitudes, and related developments in medical technologies, employment culture and so on; which were, (and still are) in turn, met with resistance and reluctance from some quarters. While this is a general example, it still refers to what is a profoundly personal experience for many. More micro-oriented approaches might be expected to be additionally concerned with the specific dynamics and processes involved in the take-up of narratives and the production of biography at any stage in the life course (e.g. Habermas and Bluck 2000; Denzin 1989). There is then movement between micro, meso and macro levels in a comprehensive narrative approach, and the way different authors and approaches tackle the relationship between these levels will reveal different balances between voluntary and socially determined action.

In sum, advocates of a narrative approach understand narrative to be a universal psychological proclivity; to have a primary and constructive role in shaping our shared understanding of reality; to therefore contribute to an ongoing sense of ontological security and to orient ourselves meaningfully in the world. We are motivated to actively develop and maintain a sense of existential coherence through the creative take-up of preexisting narrative frames. While the psychosocial proclivity for narrative organization may be universal, as may some underlying narrative (etic) structures, they are realized through culturally contingent (emic) stories—that are 'in accordance with the models of intelligibility specific to the culture' (Rosenwald 1992, p. 265). It is in terms of this psychosocial dimension—the contours of subjectivity realized through embeddedness in culturally contingent frames of meaning—that a narrative turn in the social sciences is of relevance to human action and inaction in response to knowledge of anthropogenic ecological degradation. The primacy attributed to narrative signifies how central it is in shaping our actions—our discernable behaviour as well as the way we experience and make emotional, subjective and shared sense of reality. We commonly apprehend the reality of anthropogenic ecological degradation through the existing narrative frames that constitute the societies and cultures of which we are a part. Following the premise of the narrative turn, it is clear that predominant narratives will play a key role on shaping our experience, perception and action in relation to anthropogenic ecological degradation. It is, therefore, worthwhile exploring these narratives and their relationship to anthropogenic ecological degradation in more detail.

Master Narratives and Anthropogenic Ecological Degradation

Many people report what they consider to be irrefutable and indelible impressions of ecological crisis, encroaching into their experience of everyday life, altering the very nature of their material and physical surroundings entering into their landscapes, communities, bodies; disrupting their livelihoods and life chances, threatening their physical security, and, to borrow a well-worn term form previous chapters, their ontological security (e.g., Shepherd et al. 2013; Sullivan 2014). These experiences might readily translate into stories: tales of tragedy, redemption, despair and hope. As such they potentially hold the power that narratives are claimed to have in framing reality, making truth appeals to others, and motivating them to act.

However, in relation to ecological crisis and the broader context of everyday life, 'climate change' and 'environmental problems' are not the only narratives in circulation, populating our sense-making and shared conventions. The experiences just mentioned may not be understood *as* 'climate change', as anthropogenic, or as a crisis (Lee et al. 2015). Even if they are, such experiences are embedded in wider narratives that construct individual and collective responsibility, sustainability, social and cultural norms, in ways that may or may not encourage us to 'act' in a variety of ways. These broader narratives can compliment, reject or conflict with more specific framings of ecological degradation as they intersect with them. Although we should not assume a simplistic causal relationship between narratives and actions, narratives clearly have implications for what we do—validating and legitimizing (or otherwise) an interconnected range of activities.

It follows that identifying dominant or 'master' narrative frames, not least those that underpin our collective relationship to nature, have been central to narrative scholarship. A meta- or master narrative is a coherent system of stories, explaining and organizing knowledge and experience; and rhetorically 'answering' existential questions in everyday routines, conventions, expectations and injunctions that provide everyday life in particular trajectories (Gergen 1991; Giddens 1991; Stephens and

McCallum 1998). A rich tradition of commentary and analysis in critical theory, the social sciences and beyond, points the finger at the way our relationship to the rest of the natural world is ensconced in the meta-narratives of late modernity, industrialism, capitalism and neo-liberalism (e.g., Gorz 1979; Kidner 2001; Kovel 2007).

Narratives are bound up with human knowledge, experience and activity in that they selectively confer legitimacy upon them, allow-ing the figure to emerge from the ground of undifferentiated reality. In the context of anthropogenic ecological degradation, narrations of 'nature' as inert and passive are significant, as they legitimize intensive extraction of natural resources in the pursuit of fulfilling human needs, wants and desires (Koger and Winter 2010). As a political-historical construct, 'the idea of living matter was simply economically inconve-nient [...] if nature is dead then there are no restraints on exploiting it for profit' (Wink cited in Freund 2015, p. 5). Widespread 'extractivism' is legitimized by reference to a narrative in which 'dead' or submissive nature at our disposal is juxtaposed with human mastery; and such prac-tices circularly reproduce and reinforce the 'truth' of such a narrative. The correlations here with patriarchal narratives and constructions of femininity as equivalent to nature have not been lost on ecologically-oriented feminist and critical work (e.g., Plumwood 2002; Moosa and Tuana 2014; Rust 2008).

Narrative frames come to dominate because of the way they popu-late the mundane practices and conventions of everyday life; how they engage with the contours of non-narrative reality. There is always a pos-sibility that narratives can be challenged, as we describe below, but in general terms, their power lies in the way we use them to confer a spoken legitimacy on the unspoken flow of everyday acts. Narratives of 'inert nature' are not just woven into the stories we explicitly tell ourselves about nature and our relationship to nature, but into the ways we tacitly navigate the physical world, the way we use material objects and discard them. And we do this in conjunction with others—not just the others in our immediate vicinity, but virtual, mediated and historical others, in an increasingly multiphrenic environment—one in which we are exposed to, and can often project, a plethora of mediated messages regarding self-hood (Gergen 1991). This is an environment in which the circulation of master narratives provides collective reassurance (Eden 2012). Marshall describes this power in straightforward social psychological terms:

> Such stories attain their validity through the social 'proof' of peer transmission…We follow the social cues of the people we know and trust – our friends, families and preferred media – in the selection of our preferred story of climate change. When we have views that conflict with the social norm around us we choose to suppress them rather than endanger our social allegiances. (Marshall 2014b, p. 97)

Master narratives incorporate more than an understanding of the nature of nature. Kingsnorth and Hine (2010) identify myths of progress, or perhaps more accurately progress as defined by unlimited economic growth, the array of consumer goods on offer; and anthropocentrism and dualism in terms of the idea that we, and only we, are the conscious and intelligent beings at the top of the hierarchy of life, who must seek mastery over a formidable opponent in nature (see also Klein 2011).

From a critical perspective, master narratives are often pejoratively understood as intertwined with pathological social relationships and self-understandings that are extended to human-nonhuman relationships and ecological degradation (e.g., Lasch 1979; Cushman 1990). They are culturally and historically contingent—hence attempts to trace the deeper roots of Enlightenment-based narratives and their provincial claims to universalist understandings of the meaning and purpose of human life (Mestrovic 1998). These include 'an unquestionable faith in the separateness of self and surroundings; a teleology of self-mastery; a grasping of a meaningful life as a rationally-induced future-oriented project; a disjunctive relationship between language and reality which progressive reason can overcome, or in a view of the individual as a bounded, cognitive isolate' (Adams 2003, p. 226). While valorizing 'rational consciousness' and associated mastery, industriousness, instrumentalism and dominion over nature on the one hand (Adorno and Horkheimer 2002[1944]: xi); on the other is a romanticized view of nature as sacred, a pristine paradise in retreat. In acknowledging these deeper roots, the parallel narratives of enlightenment and anti-enlightenment are said to still shape the Western worldview of nature.

While we could pursue accounts of various paradigms further, for current purposes it is sufficient to note that despite their varied expression, in most accounts these master narratives are understood to shape how we collectively understand the ecological crisis, and, vitally, how we respond to it. The 'dominant social paradigm' and related narrative frames encourage suspicion of science, rejection of dependency and interdependency and a reluctance to adopt alternative framings of environmental 'goods' (Dunlap and Liere 1984). More broadly, they position human-nonhuman interrelationship in a frame that works 'against environmentalism and against dealing with global warming' (Lakoff 2010, p. 74).

Too often the existence of various strands of a master narrative are asserted, and we can nod our heads as they elide with the reader's sense of the plots and stories they are exposed to, and which, one must assume, are weaved into their own narrative identities to various degrees. But on this point, a narrative approach must be scrutinized further. If the reader is aware of the contours of these narratives, can recognize them, and appraise them, then there must be vantage points from which they can be acknowledged as such.[4] If we pursue the 'life is narrative' claim, then this positioning must also take the narrative form, a further metanarrative within which this or that narrative is recognized as contingent, but still itself psychosocially coherent thanks to its narrative form.

Following this line of reasoning, we are not as in thrall to the master narratives described to us as the original assertion of their dominance asserts. This reinserts the possibility of agency—we have some awareness of master narratives *as* master narratives, existing on a broader plane of narrative identity. This suggests master narratives are subject to scrutiny, an incomplete in the organization of knowledge and experience (Craib 2000, 2003). It also opens up the possibility of alternative or 'new' narratives having an ontological and epistemological stake (Dunlap 2008). These alternatives may be prefigurative, more difficult to establish within and between us, but the stickiness of master narratives implied in the issue of 'difficulty' needs to be explored further and can, at least potentially, be unstuck. We return to the potential of alternative narratives below and in the following chapter. What we must consider for now is *how* master narratives are considered to achieve this hegemony and, more pragmatically, how they perpetuate denial and inaction in the face of ecological crisis. To do so, we return to the theme of consumerism, building on the analysis of previous chapters.

'It's All Folded into Normalcy': Consumerism as Narrative Vehicle for the Social Organization of Denial

The quotation in the chapter title and this subheading is taken from an interview with Bill Talen, aka environmentalist performance artist Reverend Billy of the Church of Stop Shopping. Here it is in context: 'we have an extinction wave in the Earth that the human media's ignoring. So it's the part of climate change that you don't want to talk about because it's too awful, right? [...] It's all folded into normalcy' (Talen, cited in Freund 2015, p. 10).

The target for much of Talen's critical energy is consumerism—it is a consumerist 'normal' into which the awfulness of waves of extinction and a changing climate is folded. Talen offers a contemporary version of situationism, but it has many parallels with longstanding sociological critiques of consumerism, charting its apparent ascendance as a basis for shared identity and ontological security (e.g., Campbell 2004; Bauman 2007a, b; Featherstone 2007).

A related facet of the appeal of consumerist narratives, and the legitimacy they convey, more explicitly connects it with the social organization of denial and inaction in the face of anthropogenic ecological degradation (Norgaard 2011). We here begin to build upon arguments in the previous chapter about the combined affective and social dynamics of defence mechanisms of denial. Numerous commentators point to consumerism as a pervasive narrative frame that socially organizes denial of the reality of the threat of anthropogenic ecological degradation, primarily by assuaging the uncomfortable emotions that arise from knowing about it, or more accurately, the simultaneous knowing and not-knowing characteristic of denial (Cohen 2013).

Weintrobe, as cited in the previous chapter, forthrightly states it thus: 'If we know that our actions cause damage we feel guilt and shame. However, in the mindless world of the quick fix, magical ways are found to deny guilt and shame and also to deflect us from knowing about the underlying causes of our mounting anxiety.' (Weintrobe 2010, p. 119). The 'magical ways' refers to the capacity of the imagined, anticipated or actual act of consumption to assuage uncomfortable feelings associated with affective experiences of guilt, shame and anxiety relating to the unsustainability of our actions. However, to dismiss the 'quick fix' of consumerism as 'mindless' is a disservice both to the complex ways in which a consumerist narrative is enmeshed with our everyday lives (Wilk 2010). In fact, it is commonly argued that consumerism is such a protean and powerful narrative that it its advocates manage to emplot representations of, and responses to, ecological crisis as further variants of consumerism, such as we see in corporate appropriations of sustainability and green consumption (Littler 2008; Randall 2009).

In asserting the grip of consumerism as a master narrative, there is a danger that we lose sight of the multiple narrative frames that help or hinder the ways in which different individuals, groups and communities make sense of anthropogenic ecological degradation (Lakoff 2010). Key tropes such as 'climate change' have made their way into public awareness and frame the ways in which at least some segments of those publics make sense of reality (Luke 2015). Thus 'climate change' and an array of related plot lines (mass extinction, plastic oceans, air pollution) amount to intersecting narrative frames. They might not operate on the same level as 'consumerism' but how they are appropriated by, intersects with, but also potentially challenges consumerist frames, is of vital interest. The messy business of maintaining a coherent narrative identity in everyday life involves the imperfect entanglement of a range of narratives that cannot be reduced to one master; nor can it seamlessly encompass the non-narrative reality with which it engages.

Apocalypse When?

Beyond the identification of broader cultural narratives that inhibit or encourage senscience of ecological degradation (Kingsnorth and Hine 2010), how might a proclivity for the narrative structuring of experience and identity help us to understand individual and collective responses? The emergence of narrative approaches and analysis in this area has encouraged a varied and growing body of research. There is emphasis on cycles in media coverage (McComas and Shanahan 1999); the relationship between the way people narratively frame environmental issues and environmental concern and activism (Taylor 2000); on how narrative frames shape individual and community coping strategies in the face of events associated with anthropogenic ecological crisis, such as extreme weather (e.g., Whitmarsh 2008); and on perceptions of risk in relation to climate change (e.g., Wibeck 2014). In some work, the empirical focus is the genesis of narratives within different interpretive communities (e.g., Callison 2014; Leiserowitz 2005); the importance of narrative framing to the communication and reception of climate change information (Morton et al. 2011; Spence and Pidgeon 2010); and the role of narratives in communicating future climate scenarios (e.g., Swart et al. 2004).

What unites otherwise highly varied analyses of climate change imaginaries is the argument that apocalypse narratives are counter-productive, in the sense that they do not facilitate personal responsibility and collective action, but exacerbate inertia (Hoggett 2011; Hulme 2009; Randall 2009; Swyngedouw 2010). At first glance, it might seem counter-intuitive to assert that contemplating catastrophe might serve as a collectively reassuring narrative. For Randall, mediated catastrophes are larger-than-life, extreme, and therefore, encourage a common-sense experiential incredulity in comparison to the relative imperturbable quality of everyday life. A related point is that such extremity reinforces the sense of disconnect between individual acts and enormous systemic change (Hulme 2009). When change is imagined as drastic, massive and all-at-once, the consequence of anything we do at the level of the everyday can appear anonymous and futile (Zizek 2009, p. 454).

Consider the way apocalypse narratives meld in the public consciousness with the rich seam of disaster fiction and film adds to the unreality of apocalypse, even as we entertain the idea with increasingly sophisticated and visceral technologies. In fact, the more sophisticated future scenario modelling gets, the more the parallels with entertainment industry representations, and the more seductive its properties as a fantasy. Here we see how apocalyptic imagery might be appropriated within a consumerist narrative. We consume the apocalypse, and the commodified apocalypse is a way of making it 'safe', rendering it within a constellation of meanings we already understand (vicarious thrills, box office, celebrity actors), and accordingly do not have to act out beyond the usual parameters. On this point, Timothy Luke's account of climate change as an ongoing apocalypse narrative is worth reproducing in full:

Climate change now appears to be a collectively acted, globally produced, and continuously staged new disaster movie without a single director, but with billions of producers following simple scripts. It is a 3-D film of which daily out-takes are routinely released on the evening news, Internet, and science programming, which all brim with shots of degradation, destruction, and death. Droughts, floods, wildfires, famines, heat waves are all carefully documented in this never-ending 3-D epic, but the full impact is never calculated (Luke 2015). Implicitly, anyone who thinks about the outcome of the film suspects the ending will be devastating, like *The Road Warrior, Water World, The Day After Tomorrow, The Book of Eli, The Road* or even *World War Z*, but it will not really be 'Doomsday.' Someone, somewhere, somehow, and in some fashion, will survive the droughts, floods, storms, and extinctions, and the 'sustainable development' of that fraction of humanity will then win out. In fact, the political imaginary of climate change basically exploits such mythic hopes with its futurological depictions of Greenland melting, New York under water, Omaha desertified, Florida a new archipelago or the Arctic Ocean with ice-free open seas year round. At first, it is terrifying, but the 3-D movie allows audiences to just dream up their own adventure-filled sequel, which some climate change experts actually celebrate (Luke 2015, p. 291).

The 'mythic hope' Luke describes follows the familiar contours of the redemption narrative, written into apocalyptic imaginary. The redemption narrative is argued to be a master narrative *par excellence* in modern Western cultures (McAdams 2006). As with all narratives, plot is central to creating a meaningful trajectory (Polkinghorne 1987), and the plot lines of a redemption narrative include facing substantial adversity, perhaps after an initial fall from grace or favour, facing up to the challenge, and eventually overcoming it, a trajectory that results in an enlightened or improved individual or group (McLean 2008).

Hoggett (2011) discerns 'apocalyptic survivalism', following Lasch (1984), in such narratives. Apocalyptic takes on redemption tend to reserve it for a select few, i.e., only 'a saving remnant will survive the end of the world and build a better one' (Lasch 1984, p. 83). Hoggett is accordingly critical of environmentalist narratives that engage in 'catastrophism' as they potentially displace responsibility for solutions onto a future 'elite', and write off our ability to respond effectively in the present.[5] Hoggett rides roughshod over the nuances of environmentalist engagement with an apocalyptic imaginary and their implications. The more significant implications of catastrophized future scenarios reside in their fit with the social and temporal organization of denial as described by Randall; and in the ability of consumerist narratives to appropriate catastrophe as a form of vicarious thrill.

There have been numerous accounts highlighting sophisticated narrative frames that deny the implications of ecological degradations (Cohen 2013; Dickinson 2009; Hoggett 2012; Hollander 2009; Lertzman 2010; Norgaard 2006, 2011; Opotow and Weiss 2000). All share an emphasis on how denial is propagated via, to paraphrase Berardi again (as cited in Eden 2012, p. 29), the continuous circulation of collectively reassuring narratives, which simultaneously take flight psychologically as 'internal propaganda' (Hoggett 2012). These narratives, may be routinely experienced as banal (Hilton 2008), rather than consciously constructed and reflected upon. But they are also employed artfully, thriving on 'ambiguity, illusion, evasiveness, trickery, collusion and guile' (Hoggett 2013, p. 60).

Shop Till You Drop: Consumerism, Narrative and Terror Management

Providing empirical support for claims about the grip culture-wide narratives have over what we do, and their relationship to apparent denial and inaction regarding anthropogenic ecological degradation, is notoriously difficult. One possible empirical source is the body of work carried out in establishing Terror Management Theory, discussed at length in previous chapters. To reiterate briefly, the basic premise of Terror Management Theory, following the philosophy of Ernest Becker amongst others, is that awareness of death and finitude profoundly impacts the way human beings create order and meaning in life. Much of our attempt to construct a coherent identity is haunted by the desire to overcome the reality of death; it is a perpetual defense against it.

Of course this is literally impossible, so we try to defend against death by achieving immortality symbolically, through various well-worn cultural avenues: 'The uniquely human awareness of death and the potentially overwhelming existential anxiety it engenders motivates people to imbue life with meaning and derive self-esteem from cultural beliefs about the nature of reality'. Stated simply 'we develop and maintain a solution to the problem of death by creating culture and putting faith in cultural worldviews' (Arndt et al. 2004, pp. 198–199).

More specifically, immortality striving involves a form of supra-identification with something above and beyond the limitations of our physical selves; something which derives its status and legitimacy from a contingent cultural worldview: particular lifestyles, powerful others, ingroups, worldviews that promise permanence, power, prestige and status. Other people are vital to Terror Management Theory's particular understanding of the role of narratives—it is an avowedly social psychological analysis. The contingent nature of worldviews which act as our buffer against death anxiety means that our investment in them, and the self-esteem we derive from doing so, is in constant need of replenishment. Worldviews could always be otherwise, and the existence of diverse others (especially in a multiphrenic environment) operates as a potential source of support and contention for that worldview. This fact is claimed to explain, amongst other things, the importance and intensity of ingroup allegiance and outgroup antagonism—not least when a worldview is experienced as under threat (Arndt et al. 2004; Greenberg et al. 1990; Dickinson 2009).

The legitimacy of the narratives we live by, derived from others, communicated and negotiated in multiple mediums and settings is absolutely essential to their ability to hold death at bay. The experimental dimension of Terror Management Theory lies in the manipulation of mortality salience—controlling participant's exposure to 'death primes'. This reflects the hypothesis that the more mortality becomes salient, the more we engage in hypothetically increasing compensatory activities to keep death thoughts at bay. Of particular interest are the experiments that explore the connection between exposure to fear of death and materialist values and behaviours (e.g. Choi et al. 2007; Kasser and Sheldon 2000; Mandel and Heine (1999); and related work on environmental concern and evaluations of the natural world (Fritsche and Häfner 2012; Fritsche et al. 2010; Koole and Van den Berg 2005; Vess and Arndt 2008). Although materialism is not strictly equivalent to consumerism, they substantially overlap, and most Terror Management Theory derived predictions and analysis incorporate both.

Arndt and colleagues, for example, assert that 'when consumerism is woven deeply into a cultural fabric... we should expect materialistic routes of symbolic transcendence to be pervasive in general, and especially in response to reminders of death' (Arndt et al. 2004, p. 204). They explore 'how trepidation about death (particularly at an unconscious level) instigates and sustains materialistic pursuits, as well as directs consumer choices' (2004, p. 199). A commonly cited everyday example of this connection is the marked increase in the consumption of consumer goods in the US in the months following 9/11 (Arndt et al. 2004, p. 198; Choi et al. 2007, p. 1). If our adopted worldview is a buffer between death anxiety and the everyday ability to go on, when it is threatened we seek to reestablish one or more component of it with increased vigour. The findings from studies of mortality salience and materialist tendencies generally indicate that triggering the first heightens the latter. Arndt et al.'s review of relevant studies suggest that to the extent that people adhere to beliefs, values and goals of a 'dominant consumeristic worldview' they engage in more damaging ecological behaviours, use more 'natural resources' in forest dilemma games and more generally report 'everyday lives that leave a heavier 'ecological footprint' on the planet' (Arndt et al. 2004, p. 209; see also Kasser and Sheldon 2000; Kasser 2002; Brown and Kasser 2005).

Socially Organized Terror?

The broader implication of terror management analysis in relation to narrative approaches needs to be understood in the context of assertions that we live in an *increasingly* consumerist and materialist society (Twenge and Kasser 2013; Bauman 2007a). In 'hyper' or 'turbo' consumerism, marked by an increase in the volume and diversity of goods and services available to expanding numbers (Binkley and Littler 2008; Lawson 2009), we potentially see an accompanying rise in the number of contingent situations and dispositions in which self-esteem is invested in consumerist worldviews; the failure to able to do so perceived socially as a personal lack or flaw (Bauman 2007b); and defences of that worldview accordingly prominent (Dickinson 2009). The tragic irony of a Terror Management analysis is that as time passes, the more anthropogenic ecological degradation is communicated and experienced as a threat, i.e., mortality salient, the more vociferously an already hegemonic consumerist worldview will be defended, which further exacerbates ecological degradation.

However, Terror Management Theory predictions are complicated by salient dispositional and situational factors. These include the contingent domains in which we come to invest self-esteem, such as 'appearance, approval, competence, and virtue' (Choi et al. 2007, p. 7). For those who identify as environmentally conscientious, and attach self-esteem to those values, for example, mortality salience *increases* self-reported 'concern for the environment', while for those who do not appear to derive any self-esteem from environmental causes, mortality salience *decreases* concern for the environment (Vess and Arndt 2008).

As discussed in Chapter 6, a consideration of the wider social context, and the ways in which defences are socially organized, raises further questions about the universality of terror management dynamics. Cultures can involve narratives that contain, address and process finitude, mortality and generative capacities in a host of different ways. Social and cultural forms can potentially prescribe, contain and process how we make sense of death and mortality. Different religious philosophies perhaps most obviously point to fundamental distinctions in how death is conceptualized and experienced, but such philosophies always exist in conjunction with other historical traditions and worldviews (Park 2005).

Despite these criticisms and caveats, Terror Management Theory provides some useful linkages between differing accounts of inaction and unsustainable behaviour in the context of anthropogenic ecological degradation. The psychological concepts of denial and other defence mechanisms can be combined with more sociological emphasis on emotion, shared narrative frames and identities to articulate the psychosocial organization of defence mechanisms. A Terror Management perspective can contribute to the work of Norgaard and a handful of others who have begun to unpack the specifics of the interrelationship between affective experience, interpersonal dynamics and narrative. It provides a further account of what might motivate the adoption of defence mechanisms and their outline form in response to anthropogenic ecological degradation.

While these bodies of work emerge from different philosophical traditions, i.e., existentialism, psychoanalysis and the sociology of emotions, their insights are largely complimentary. In fact, Terror Management Theory uses the language of defences as we saw in the 'dual defence model' of terror management (Pyszczynski et al. 1999). The point here is that worldviews, narratives and the active emplotment of identity provide the constellations via which these defences are navigated, and out of which a narrative identity is assembled. The extent of our psychosocial embeddedness in, and embodiment of, these narratives from early processes of socialization and attachment onwards (Mikulincer and Florian 2000), indicates the challenge that faces those consciously aiming to building alternative narrative frames (Lakoff 2010).

Experience Exceeds Narrative

> What succor, what consolation is there in truth, compared to a story? What good is truth, at midnight, in the dark, when the wind is roaring like a bear in the chimney? When the lightning strikes shadows on the bedroom wall and the rain taps at the window with its long fingernails? No. When fear and cold make a statue of you in your bed, don't expect hard-boned and fleshless truth to come running to your aid. What you need are the plump comforts of a story. The soothing, rocking safety of a lie. (Diane Setterfield, *The Thirteenth Tale*)

With a novelist's flourish, Setterfield captures the psychological appeal of a narrative—the coherence and reassurance it can provide. She also articulates something else about stories—that the comfort they provide is a defence, a lie to hide the truth of a harsher reality. This is a thorny issue for narrative research and theory, or at least it should be. Always sceptical of the 'turn to narrative' in the social sciences, Ian Craib consistently argued that we should never lose sight of the fact that narratives are impositions on a non-narrative reality: 'there is a rooting of our narratives, even if only minimally, in a reality external to the narrative' (2002, p. 65). The reality that exceeds and is prior to narrative involves material form and relationships. How we articulate those forms, actions and interactions of course depends on the vocabulary available to us; and that repertoire is socially, culturally, historically contingent. But account for reality we must, and in the context of anthropogenic ecological degradation, it seems imperative to explicitly advocate a critical realist position (Bhaskar 2010; Crist 2004; Huckle 2004). Doing so involves being a little more explicit about the relationship between narrative and non-narrative reality, which I will attempt to do briefly here.

To some extent preceding discussion of socially organized defence mechanisms relies on implicit claims about non-narrative reality. Following the psychoanalytic tradition, unconscious affective experiences and the defence mechanisms that process them constitute that reality. In a sense the 'mechanism' in question is about the conversion of non-narrative psychic reality into narrative form. In the context of ecological degradation, non-narrative reality is the reality of ecological interdependence and anthropogenic degradation—the very material destruction and devastation that is already occurring beyond narrative; but also further predicted changes to come (warming, extinction, erosion etc.). There is a reality, of interdependent planetary life, even if we are doomed to know it partially and distortedly through our very human mix of cultural, biological, psychological and social perception—we are still, nonetheless *of* that reality, and capable of feeling it, sharing it, intuiting it and, importantly, of making claims about how well narratives represent the truth of it.

Craib adopts Sartre's concept of 'bad faith'—the disavowal of the freedom and responsibility to act—to explore how the narratives we use 'make things normal, unproblematic with the added advantage of avoiding acknowledgement or feeling guilt. At the very least they keep the problems within a recognizable plot... in this sense all personal narratives are to some degree bad faith narratives' (2002, p. 67). Perhaps so, but bad faith narratives at least point to the possibility of alternative possibilities—of problematizing 'normal' rather than folding troublesome experience into it. As a form of politics, we must also use narrative to frame how this reality is changing, under the influence of human activity, in a direction that is and will reduce the diversity of life and the ability of many forms of life to flourish, including humans. Following this logic, we can also try to acknowledge reality more faithfully, however problematic—in this context the reality of ecological degradation; and articulate alternative 'good faith narratives'—however unfinished and uncertain.

Looking for Alternative Narratives

It is difficult to sum up the importance attributed to narrative in relation to anthropogenic ecological degradation as work in this 'area' is so varied and complex. If we can pull out a few key issues at this stage, the first is the argument that the narratives we live by do have a *causal* role in initiating, maintaining and obscuring anthropogenic ecological degradation. Narratives that frame nature as an inert set of resources, for example, or humanity as at the top of a hierarchy of life, though oversimplified examples, are seen as perpetuating and legitimizing ecological degradation. A causal role might also be associated with narratives that promote inaction in the face of knowledge of anthropogenic ecological degradation. Here, we connect back to the social organization of defence mechanisms, and in this context those mechanisms take on a narrative form more explicitly. As Marshall concisely states it, 'Many of the ways that we talk about climate change are more subtle narrative constructions designed to avoid anxiety and personal responsibility' (Marshall 2014a, p. 97).

Narrative frames run deep (Lakoff 2010). For alternatives to take hold, they must muscle in on the territory of existing dominant narratives; challenging their power to construct and direct our shared understanding of reality; our ongoing sense of ontological security and our meaningful orientation in the world. How does this happen, when experiences that do not conform to emerging expectations of what a story should look and sound like are prone to be experienced as uncanny or odd, and reformulated to fit with canonical narratives (McAdams 2005, p. 242)? Does it mean that narrative deviations are possible, but must also, therefore, take recognizable narrative form (Fivush 2010)?

The hope invested in alternative narratives is the basis for transition towards genuinely sustainable societies, while acknowledging that other factors contribute to the meaningful take-up of narratives. Subsequent questions focus on which narratives, according to whom, and how they emerge and take hold. This is rich and varied territory that takes us beyond the remit of this chapter, and on to the final chapter and beyond. An interest in alternative narratives has various directions of travel theorizing and researching the conditions (material, social, cultural, political, affective, interpersonal) in which alternative narratives might or do take hold and flourish; actively encouraging and facilitating the fermentation and development of alternatives through research; identifying, studying, interpreting, developing and circulating promising alternatives. Whichever route is taken, the 'alternatives' in question may be partial or prefigurative—this is a corollary of the dominance of existing narrative frames.

Notes

1. For example the 'Seven Dimensions of Climate Change' project commissioned by the Royal Society for the encouragement of Arts, Manufactures and Commerce (RSA) is an attempt to reframe public debates about climate change https://www.thersa.org/discover/publications-and-articles/reports/the-seven-dimensions-of-climate-change-introducing-a-new-way-to-think-talk-and-act/; the Climate Outreach and Information Network (COIN) has consistently focussed on the importance of narrative in communicating climate change effectively http://www.climateoutreach.org.uk/about/; Mediating Change is an Open University research centre acting as an umbrella for various climate change and culture programmes of work that have explored the importance of narrative in a series of events, podcasts and publications http://www.open.ac.uk/researchcentres/osrc/research/themes/mediating-change

2. Lakoff argues that 'framing' is essential to narrative. Frames are typically unconscious structures that organize thinking. Whilst we might be tempted to think of frames as metaphorical rather than material, cognitive and brain scientists claim that they have an identifiable biological base: 'physically realized in neural circuits in the brain' (Lakoff 2010, p. 71). Frames allow us to discern roles (a suitable metaphor for an emphasis on narrative), the associations between them, and, according to Lakoff, the relations between different frames too. He offers the example of a hospital: 'A hospital frame, for example, includes the roles: Doctor, Nurse, Patient, Visitor, Receptionist, Operating Room, Recovery Room, Scalpel, etc. Among the relations are specifications of what happens in a hospital, e.g. Doctors operate on Patients in Operating Rooms with Scalpels' (p. 71). If frames can also indicate the

relationship to other frames, a hospital frame also slots intersects with other frames such as Welfare, Health, Work, Death and Birth. Narrative, on the other hand, is a 'system of frames': a meaningful combination of frames that communicates something as truthful (Lakoff 2010, p. 73). Lakoff does not offer any definition of narrative beyond this assertion, and is unclear about where a 'frame' ends and a 'narrative' begins.

3. The authors differentiate between two different kinds of strategy for maintaining personal persistence whilst accommodating the need for change—Narrative and Essentialist. They find that these strategies are largely contingent on culture: the first distinctly characterizes Aboriginal, the second non-Aboriginal youth. Despite their labels, both strategies rely on narrative structures in different ways to 'manage' the need for both continuity and change.

4. I am not going to entertain the divisive, elitist assumption that whilst we see through master narratives of, say consumerism, it is others—one version or another of 'the masses' – who are unwittingly transfixed.

5. Hoggett also dismisses the Dark Mountain Project for advocating a version of apocalyptic survivalism. He claims that they 'have already abandoned hope and seem to be engaged in the same kind of retreat' (2011, p. 265) and harbour the deluded belief that they alone 'have the foresight to prepare for the worst and the moral fibre to prevail' (2011, p. 268). I discuss this accusation in more detail elsewhere (Adams 2014).

References

Adams, M. (2003). The reflexive self and culture: A critique. *The British Journal of Sociology, 54*(2), 221–238.

Adorno, T. W. & Horkheimer, M. (2002 [1944]). *Dialectic of enlightenment* (trans: Jephcott, E). Stanford: Stanford UP.

Arndt, J., Solomon, S., Kasser, T., & Sheldon, K. M. (2004). The urge to splurge: A terror management account of materialism and consumer behavior. *Journal of Consumer Psychology, 14*(3), 198–212.

Atkinson, P. (1997). Narrative turn or blind alley? *Qualitative Health Research, 7*(3), 325–344.

Bauman, Z. (2007a). *Consuming life*. London: Wiley.

Bauman, Z. (2007b). Collateral casualties of consumerism. *Journal of Consumer Culture, 7*(1), 25–56.

Berger, R. J., & Quinney, R. (Eds.). (2005). *Storytelling sociology: Narrative as social inquiry*. Boulder: Lynne Rienner Publishers.

Bhaskar, R. (Ed.). (2010). *Interdisciplinarity and climate change: Transforming knowledge and practice for our global future*. London: Taylor & Francis.

Binkley, S., & Littler, J. (2008). Introduction: Cultural studies and anti-consumerism: A critical encounter. *Cultural Studies, 22*(5), 519–530.

Brown, K. W., & Kasser, T. (2005). Are psychological and ecological well-being compatible? The role of values, mindfulness, and lifestyle. *Social Indicators Research, 74*(2), 349–368.

Brown, B., Nolan, P., Crawford, P., & Lewis, A. (1996). Interaction, language and the "narrative turn" in psychotherapy and psychiatry. *Social Science and Medicine, 43*(11), 1569–1578.

Bruner, J. S. (1987). Life as narrative. *Social Research, 54*, 11–32.

Bruner, J. S. (1990). *Acts of meaning*. Cambridge, MA: Harvard University Press.

Callison, C. (2014). *How climate change comes to matter. The communal life of facts*. Durham: Duke University Press.

Campbell, C. (2004). I shop therefore I know that I am: The metaphysical basis of modern consumerism. In K. Ekstrom & H. Brembeck (Eds.), *Elusive consumption: Tracking new research perspectives* (pp. 27–44). Oxford: Berg.

Chandler, M. J., Lalonde, C. E., Sokol, B. W., Hallett, D., & Marcia, J. E. (2003). Personal persistence, identity development, and suicide: A study of native and non-native North American adolescents. In *Monographs of the society for research in child development*. Ann Arbor: SRCD.

Chandler, M. J., Lalonde, C. E., & Teucher, U. (2004). Culture, continuity, and the limits of narrativity: A comparison of the self-narratives of native and non-native youth. In C. Daiute & C. Lightfoot (Eds.), *Narrative analysis: Studying the development of individuals in society* (pp. 245–265). Sage: Thousand Oaks.

Choi, J., Kwon, K. N., & Lee, M. (2007). Understanding materialistic consumption: A terror management perspective. *Journal of Research for Consumers, 13*, 1.

Cohen, S. (2001). *States of denial: Knowing about atrocities and suffering*. New York: Wiley.

Cohen, S. (2013). Discussion. In S. Weintrobe (Ed.), *Engaging with climate change: Psychoanalytic and interdiscipinary perspectives* (pp. 72–79). London: Routledge.

Craib, I. (2000). Narratives as bad faith. In M. Andrews, S. D. Sclater, C. Squire, & A. Treacher (Eds.), *Lines of narrative: Psychosocial perspectives* (pp. 64–74). London: Routledge.

Craib, I. (2003). The unhealthy underside of narratives. In C. Horrocks (Ed.), *Narrative, memory and health* (pp. 1–11). Huddersfield: University of Huddersfield.

Crist, E. (2004). Against the social construction of nature and wilderness. *Environmental Ethics, 26*, 5–24.

Cushman, P. (1990). Why the self is empty: Toward a historically situated psychology. *American Psychologist, 45*(5), 599.

Czarniawska, B. (2004). The 'narrative turn' in social studies. In *Narratives in social science research. Introducing qualitative methods*. London: Sage.

Denzin, N. (1989). *Interpretive biography*. London: Sage.

Dickinson, J. L. (2009). The people paradox: Self-esteem striving, immortality ideologies, and human response to climate change. *Ecology and Society, 14*(1), 34.

Du Cann, C., Hunt, N., Strang, E., & Wheeler, S. (Eds.) (2015). *Dark Mountain Issue 7*. Dark Mountain Project.

Dunlap, R. E. (2008). The new environmental paradigm scale: From marginality to worldwide use. *Journal of Environmental Education, 40*(1), 3–18.

Dunlap, R. E., & Liere, K. D. (1984). Commitment to the dominant social paradigm and concern for environmental quality. *Social Science Quarterly, 65*(4), 1013–1029.

Eden, D. (2012). Angels of love in the unhappiness factory. *Subjectivity, 5*(1), 15–35.

Featherstone, M. (2007). *Consumer culture and postmodernism* (2 ed.). Los Angeles: Sage.

Fisher, W. R. (1987). *Human communication as narration: Toward a philosophy of reason, value, and action* (Vol. 201). Columbia: University of South Carolina Press.

Fivush, R. (2010). Speaking silence: The social construction of silence in autobiographical and cultural narratives. *Memory, 18*(2), 88–98.

Freund, J. (2015). Rev Billy vs. the Market: A sane man in a world of omnipotent fantasies. *Journal of Marketing Management, 31*(13–14), 1529–1551.

Fritsche, I., Jonas, E., Kayser, D. N., & Koranyi, N. (2010). Existential threat and compliance with pro environmental norms. *Journal of Environmental Psychology, 30*(1), 67–79.

Fritsche, I., & Häfner, K. (2012). The malicious effects of existential threat on motivation to protect the natural environment and the role of environmental identity as a moderator. *Environment and Behavior, 44*(4), 570–590.

Gergen, K. (1991). *The saturated self: Dilemmas of identity in contemporary life*. New York: Basic books.

Gergen, K. J., & Gergen, M. M. (1988). Narrative and the self as relationship. In L. Berkowitz (Ed.), *Advances in experimental social psychology* (pp. 17–56). New York: Academic Press.

Gergen, K. J. (2005). Narrative, moral identity, and historical consciousness. In J. Straub (Ed.), *Narration, identity and historical consciousness* (pp. 99–119). Oxford: Berghahn.

Gergen, K., & Gergen, M. (1986). Narrative form and the construction of psychological science. In T. R. Sarbin (Ed.), *Narrative psychology: The stories nature of human conduct*. New York: Praeger.

Giddens, A. (1991). *Modernity and self-identity*. Cambridge: Polity Press.

Gjerde, P. F. (2004). Culture, power, and experience: Toward a person-centered cultural psychology. *Human Development, 47*, 138–157.

Gorz, A. (1979). *Ecology as politics*. London: Southend Press.

Graugaard, J. (2014). *Transforming sustainabilities: Grassroots narratives in an age of transition. An ethnography of the Dark Mountain Project*. Unpublished PhD manuscript, University of East Anglia. http://refiguring.net/thesis.html

Gray, J. (2009, September 10). Uncivilisation: The Dark Mountain Manifesto. *New Statesman*. http://www.newstatesman.com/books/2009/09/civilisation-planet-authors. Accessed 18 Dec 2015.

Greenberg, J., Pyszczynski, T., Solomon, S., Rosenblatt, A., Veeder, M., Kirkland, S., & Lyon, D. (1990). Evidence for terror management 11: The effects of mortality salience on reactions to those who threaten or bolster the cultural worldview. *Journal of Personality and Social Psychology, 58*, 308–318.

Habermas, T., & Bluck, S. (2000). Getting a life: The emergence of the life story in adolescence. *Psychological Bulletin, 126*, 748–769.

Hilton, M. (2008). The banality of consumption. *Citizenship and consumption* (pp. 87–103). Basingstoke: Palgrave.

Hoggett, P. (2011). Climate change and the apocalyptic imagination. *Psychoanalysis, Culture and Society, 16*(3), 261–275.

Hoggett, P. (2013). Climate change in a perverse culture. In S. Weintrobe (Ed.), *Engaging with climate change: Psychoanalytic and interdisciplinary perspectives* (pp. 56–71). London: Routledge.

Hollander, N. C. (2009). When not knowing allies with destructiveness: Global warning and psychoanalytic ethical non-neutrality. *International Journal of Applied Psychoanalytic Studies, 6*(1), 1–11.

Huckle, J. (2004). Critical realism: A philosophical framework for higher education for sustainability. In P. B. Corcoran & A. E. Wals (Eds.), *Higher education and the challenge of sustainability* (pp. 33–47). Berlin: Springer.

Hulme, M. (2009). *Why we disagree about climate change: Understanding controversy, inaction and opportunity.* Cambridge: Cambridge University Press.

Hunt, N., Kingsnorth, P., & Wheeler, S. (2014). *Dark mountain issue 6.* Ulverston: Dark Mountain Project.

Josselson, R., & Lieblich, A. (Eds.). (1999). *Making meanings of narratives. The narrative study of lives.* London: Sage.

Kasser, T., & Sheldon, K. M. (2000). Of wealth and death: Materialism, mortality salience, and consumption behavior. *Psychological Science, 11*, 348–351.

Kasser, T. (2002). *The high price of materialism.* Cambridge: MIT press.

Kidner, D. W. (2001c). *Nature and psyche: Radical environmentalism and the politics of subjectivity.* Albany: SUNY Press.

Kingsnorth, P., & Hine, D. (Eds.). (2010a). *Dark mountain issue 1.* Ulverston: Dark Mountain Project.

Kingsnorth, P., & Hine, D. (Eds.). (2010b). *Editorial: It's the end of the world as we know it (and we feel fine). In: Dark mountain issue 1* (pp. 1–5). Ulverston: Dark Mountain Project.

Kingsnorth, P., & Hine, D. (2011). Editorial: Control and other illusions. In P. Kingsnorth & D. Hine (Eds.), *Dark mountain issue 2* (pp. 1–3). Dark Mountain Project: Ulverston.

Klein, N. (2011). On precaution. In P. Kingsnorth & D. Hine (Eds.), *Dark Mountain Issue 2* (pp. 20–25). Dark Mountain Project: Ulverston.

Koger, S. M., & Winter, D. D. (2010). *The psychology of environmental problems: Psychology for sustainability.* Hove: Psychology press.

Koole, S., & Van den Berg, A. (2005). Lost in the wilderness: Terror management, action-orientation, and nature evaluation. *Journal of Personality and Social Psychology, 88*, 1014–1028.

Kovel, J. (2007). *The enemy of nature: The end of capitalism or the end of the world?* New York: Zed Books Ltd.

Lakoff, G. (2010). Why it matters how we frame the environment. *Environmental Communication, 4*(1), 70–81.

Lasch, C. (1979). *The culture of narcissism.* London: Norton.

Lasch, C. (1984). *The minimal self: Psychic survival in troubled times.* London: Norton.

Lawson, N. (2009). *All consuming.* Harmondsworth: Penguin.

Lee, T. M., Markowitz, E. M., Howe, P. D., Ko, C. Y., & Leiserowitz, A. A. (2015). Predictors of public climate change awareness and risk perception around the world. *Nature Climate Change, 5*, 1014–1020.

Leiserowitz, A. A. (2005). American risk perceptions: Is climate change dangerous? *Risk Analysis, 25*(6), 1433–1442.

Lertzman, R. (2010). Desire, longing and the return to the garden: Reflections on *avatar*. *Ecopsychology, 2*(1), 41–43.

Littler, J. (2008). *Radical consumption: Shopping for change in contemporary culture*. London: McGraw-Hill.

Luke, T. W. (2015). The climate change imaginary. *Current Sociology, 63*(2), 280–296.

Mandel, N., & Heine, S. J. (1999). Terror management and marketing: He who dies with the most toys wins. *Advances in Consumer Research, 26*, 527–532.

Mandler, J. M. (1984). *Stories, scripts, and scenes: Aspects of schema theory*. Hillsdale: Erlbaum.

Marshall, G. (2014a). *Don't even think about it: Why our brains are wired to ignore climate change*. London: Bloomsbury.

Marshall, G. (2014b). Five. In J. Smith, R. Tyszczuk, & R. Butler (Ed.), *Culture and climate change: Narratives* (Vol. 2, pp. 96–97). Cambridge: Shed.

McAdams, D. P. (1993). *The stories we live by*. New York: Guilford.

McAdams, D. P. (2005). Studying lives in time: A narrative approach. *Towards an Interdisciplinary Perspective on the Life Course Advances in Life Course Research, 10*, 237–258.

McAdams, D. P. (2006). The redemptive self: Generativity and the stories Americans live by. *Research in Human Development, 3*(2–3), 81–100.

McComas, K., & Shanahan, J. (1999). Telling stories about global climate change measuring the impact of narratives on issue cycles. *Communication Research, 26*(1), 30–57.

McLean, K. C. (2008). The emergence of narrative identity. *Social and Personality Psychology Compass, 2*(4), 1685–1702.

McNay, L. (1999). Gender, habitus and the field: Pierre Bourdieu and the limits of reflexivity. *Theory Culture and Society, 16*(1), 95–117.

Mestrovic, S. G. (1998). *Anthony Giddens: The last modernist*. London: Routledge.

Mikulincer, M., & Florian, V. (2000). Exploring individual differences in reactions to mortality salience: Does attachment style regulate terror management mechanisms? *Journal of Personality and Social Psychology, 79*, 260–273.

Monbiot, G. (2010, May 10). I share their despair, but I'm not quite ready to climb the Dark Mountain. *The Guardian*. http://www.theguardian.com/commentisfree/cif-green/2010/may/10/deepwater-horizon-greens-collapse-civilisation. Accessed 18 Dec 2015.

Moosa, C. S., & Tuana, N. (2014). Mapping a research agenda concerning gender and climate change: A review of the literature. *Hypatia, 29*(3), 677–694.

Morton, T. A., Rabinovich, A., Marshall, D., & Bretschneider, P. (2011). The future that may (or may not) come: How framing changes responses to uncertainty in climate change communications. *Global Environmental Change, 21*(1), 103–109.

Norgaard, K. M. (2006). "We don't really want to know": Environmental justice and socially organized denial of global warming in Norway. *Organization and Environment, 19*(3), 347–370.

Norgaard, K. M. (2011d). *Living in denial: Climate change, emotions and everyday life.* Cambridge: MIT Press.

Opotow, S., & Weiss, L. (2000). New ways of thinking about environmentalism: Denial and the process of moral exclusion in environmental conflict. *Journal of Social Issues, 56*(3), 475–490.

Park, C. L. (2005). Religion as a meaning-making framework in coping with life stress. *Journal of Social Issues, 61*(4), 707–729.

Plumwood, V. (2002). *Environmental culture: The ecological crisis of reason.* London: Routledge.

Polkinghorne, D. E. (1987). *Narrative knowing and the human sciences.* New York: SUNY Press.

Pyszczynski, T., Greenberg, J., & Solomon, S. (1999). A dual-process model of defense against conscious and unconscious death-related thoughts: An extension of terror management theory. *Psychological Review, 106*, 835–845.

Randall, R. (2009). Loss and climate change: The cost of parallel narratives. *Ecopsychology, 1*(3), 118–129.

Ricoeur, P. (1984). *Time and narrative.* Chicago: University of Chicago Press.

Ricoeur, P. (1985). History as narrative and practice. *Philosophy Today, 29*(3), 213–222.

Rosenwald, G. (1992). Conclusion: Reflections on narrative self-understanding. In G. Rosenwald & R. L. Ochberg (Eds.), *Storied lives: The cultural politics of self-understanding* (pp. 265–289). New Haven: Yale University Press.

Rust, M. J. (2008). Climate on the couch: Unconscious processes in relation to our environmental crisis. *Psychotherapy and Politics International, 6*(3), 157–170.

Sarbin, T. R. (Ed.). (1986). *Narrative psychology: The storied nature of human conduct.* New York: Prager.

Shepherd, A., Mitchell, T., Lewis, K., Lenhardt, A., Jones, L., Scott, L., & Muir-Wood, R. (2013). *The geography of poverty, disasters and climate extremes in 2030.* London: ODI.

Smith, D. (2014a, April 17). It's the end of the world as we know it…and he feels fine. *New York Times.* http://www.nytimes.com/2014/04/20/magazine/

its-the-end-of-the-world-as-we-know-it-and-he-feels-fine.html?_r=0. Accessed 18 Dec 2015.

Smith, J. (2014b). Counter-hegemonic networks and the transformation of global climate politics: Rethinking movement-state relations. *Global Discourse, 4*(2–3), 120–138.

Spence, A., & Pidgeon, N. (2010). Framing and communicating climate change: The effects of distance and outcome frame manipulations. *Global Environmental Change, 20*(4), 656–667.

Stephens, J., & McCallum, R. (1998). *Retelling stories, framing culture: Traditional story and metanarratives in children's literature*. New York: Taylor & Francis.

Sullivan, R. (Ed.) (2014). *Moving stories. The voices of people who move in the context of environmental change*. London: Climate Outreach Network. http://climatemigration.org.uk/moving-stories-report-the-voices-of-people-who-move-in-the-context-of-environmental/. Accessed 21 Dec 2015.

Swart, R. J., Raskin, P., & Robinson, J. (2004). The problem of the future: Sustainability science and scenario analysis. *Global Environmental Change, 14*(2), 137–146.

Swyngedouw, E. (2010). Apocalypse forever? Post-political populism and the spectre of climate change. *Theory, Culture & Society, 27*(2-3), 213–232.

Taylor, D. E. (2000). The rise of the environmental justice paradigm injustice framing and the social construction of environmental discourses. *American Behavioral Scientist, 43*(4), 508–580.

Townsend, S. (2010, June 3). Get down off your Dark Mountain: You're making matters worse. *The Ecologist*. http://www.theecologist.org/blogs_and_comments/commentators/other_comments/498336/get_down_off_your_dark_mountain_youre_making_matters_worse.html. Accessed 18 Dec 2015.

Twenge, J. M., & Kasser, T. (2013). Generational changes in materialism and work centrality, 1976–2007: Associations with temporal changes in societal insecurity and materialistic role modeling. *Personality and Social Psychology Bulletin, 39*(7), 883–897.

Vess, M., & Arndt, J. (2008). The nature of death and the death of nature: The impact of mortality salience on environmental concern. *Journal of Research in Personality, 42*, 1376–1380.

Weintrobe, S. (2010). Engaging with climate change means engaging with our human nature. *Ecopsychology, 2*(2), 119–120.

Whitmarsh, L. (2008). Are flood victims more concerned about climate change than other people? The role of direct experience in risk perception and behavioural response. Tyndall Centre for Climate Change Research, School of Psychology Cardiff University. http://psych.cf.ac.uk/home2/whitmarsh/ Whitmarsh%20J%20of%20Risk%20Research%202008.pdf. Accessed 19 Sep 2016.

Wibeck, V. (2014). Social representations of climate change in Swedish lay focus groups: Local or distant, gradual or catastrophic? *Public Understanding of Science, 23*(2), 204–219.

Widdershoven, G. A. (1993). The story of life: Hermeneutic perspectives on the relationship between narrative and life history. *The Narrative Study of Lives, 1*, 1–20.

Wilk, R. (2010). Consumption embedded in culture and language: Implications for finding sustainability. *Sustainability: Science, Practice, & Policy, 6*(2), 38–48.

Yuval-Davis, N. (2006). Intersectionality and feminist politics. *European Journal of Women's Studies, 13*(3), 193–209.

Žižek, S. (2009). *In defense of lost causes*. London: Verso.

10

Embodied Entanglements: Exploring Trans-Species

Introduction

> Psychology has become trans-species, and in so doing has transformed talking with animals from the fantastic to the commonsensical. (Gay Bradshaw)

The late ecofeminist philosopher Val Plumwood identified an 'ecological crisis of reason' (2002) in which 'the dominant rationalist paradigm overwrites the emotional, experiential and embodied entanglements of humans and their environments' (cited in Paschen and Ison 2014, p. 1087). In one sense, this statement chimes with explanations of ecological destruction and inaction in terms of narrative frames—'the dominant rationalist paradigm' explored in earlier chapters. However, she also asserts that this narrative 'overwrites' something other-than-narrative, something tangible and material—'emotional, experiential and embodied entanglements of humans and their environments'.

In the context of ecology and ecological crisis, entanglement refers, in plain language, to our (meaning human) relationship with

209

M. Adams, *Ecological Crisis, Sustainability and the Psychosocial Subject*,
DOI 10.1057/978-1-137-35160-9_10

more-than-human nature. Ecology is equated with the interdependence of life, and ecological crisis with a catastrophic threat to that interdependence. What is being overwritten here is not the inherent interdependence of life *per se*, but the sociocultural and psychological recognition and validation of interdependence, and, presumably, opportunities for experiences of it free from distortion. Interdependence is still claimed to exist, although it is 'hidden' beneath the inscriptions of rationalism, industrialism, consumerism, or amidst them, if most likely, in opaque or distorted forms.

For the most part, mainstream social science concerned with anthropogenic ecological degradation does not yet dwell on human and more-than-human entanglements.[1] Outside the mainstream, there are numerous critical and interdisciplinary developments that *do* try to provide some kind of analytical framework that can incorporate the human and more-than-human word and their intertwinement (Choy et al. 2009; Deleuze and Guattari; Kirksey and Helmreich 2010; Haraway 2008; Cassidy and Mullin 2007). Such ventures are many and varied, including, for example, the 'growing currency' of posthumanist discourse and critical engagements with it (e.g. Haraway 2003; Hird 2010; Pickering 2005; Wolfe 2010), some developments in psychology, particularly trans-species psychology and radical ecopsychology (e.g. Bradshaw and Watkins 2006; Fisher 2013a, b) and social theory (e.g. Kidner 2001, 2012; Latour 1993, 2005).

Of particular interest are studies that allow us to pull back from what can feel like the 'generic anonymity' of 'the environment' in 'pro-environmental behaviour'; and simultaneously challenge a pervasive sense of dissociation and independence from the more-than-human world that the organization of contemporary societies often encourages (Franklin 1999; Worthy 2013). Constructions of 'the environment' often reach past the messier business of phenomenal relationships between the human and more-than-human world, to an abstracted plane of interaction that seems to float above reality ('human' and 'environment'). To pursue this enquiry further, in this chapter we will focus in on human-nonhuman *animal* interaction, as it offers a particular manifestation of Plumwood's broader sense of 'entanglement'.

Trans-Species Psychology and Speciesism

If David Kidner could attest at the beginning of the twenty-first century that the 'discontinuity between the 'animal' and 'human' realms is beginning to come under fire' (2001, p. 94), the intervening years have seen a gathering of pace. Across numerous disciplines, there is growing recognition of species interdependence, as Haraway evocatively asserts here: 'There has been an explosion within the biologies of multi-species becoming with; understanding that to be a one at all you have to be a one of many, and that is not a metaphor; it is about the tissues of being anything at all; and that those who are have been in relationality all the way down' (Haraway 2014). Such a challenge simultaneously encourages us to consider the ethical and political dimensions of our relationship with nonhuman animals, not least as they are inscribed in social, cultural (including narrative) and material conventions (Cudworth 2015). Trans-species psychology is one area that explicitly tries to address this challenge.

To define it at its broadest, trans-species psychology is 'the formal study of how animals think, feel and behave' (Bradshaw 2010, p. 158). While this might appear uncontroversial, the novelty in this definition is that 'animals' most definitely includes humans. The prefix 'trans' is there to suggest that 'a common model of psyche applies for all human species, including humans' (Bradshaw 2010, p. 158). This common model is utilized to inform a number of theoretical developments, campaigns and research programmes, emphasizing the complexity of animal experience, points of continuity and difference between humans and other species, and identifying interspecies interdependence (e.g. Bradshaw et al. 2008; Buckley and Bradshaw 2010). However, a growing recognition of 'becoming with' across numerous disciplines should not be taken to suggest that the path is, therefore, clear for more ethical practices that cherish interdependence and mutual stakes in relationality. Humans are *already* entangled with the existence of other species, and the predominant forms of interrelationship make, and are made possible by, existing social, material, historical, cultural, political and psychological practices.

A starting point for exploring the predominant forms these practices make is the concept of speciesism—'the belief in the inherent superiority of one species over others' (Moore 2013, p. 12); a concept first explicitly advanced by the British psychologist and animal rights activist Richard Ryder (1975). Trans-species psychologists consider speciesism to be 'a primary cultural organizing principle', deeply inscribed in society, culture and psyche (Bradshaw and Watkins 2006, p. 6; Potts 2010). Bradshaw and Watkins go on to state 'human-animal differencing comprises much of what defines western human collective identity and an ego construct based on what animals are presumed to lack' (Bradshaw and Watkins 2006, p. 7).

In a related discussion, Cudworth makes a specific call to incorporate 'systematic cruelty' towards other species into contemporary sociologies of violence (2015), echoing the longstanding arguments of many eco-feminists, animal rights activists and environmentalists (Harrison 1964; Collard 1988; Adams 2015a). As Cudworth attests, 'the statistics are of staggering proportions', particularly in relation to animal agriculture: at least 55 billion land-based nonhuman animals are killed in the farming industry per year, and this figure is growing year-on-year. The domain of companion animals is relatively small, but 3–4 million stray dogs and cats are killed a year in the USA alone (Cudworth 2015). In Cudworth's approach, such violence reflects the complex intersections of relations of social power, embedded in everyday practices and discourses to the extent that it is a 'routine, unexceptional and part of the fabric of 'British life' (Cudworth 2015, p. 14). She argues that while forms of contestation do exist, they run parallel to the perpetuation and growth of systematized violence on an unprecedented scale:

> Indeed, the articulation of political claims on behalf 'of' animals has proceeded alongside the global spread of Western intensive animal agriculture and an enormous increase in animal populations bred and killed for food. Where the lives of domesticated nonhuman animals are concerned, Bauman's (1989) thesis is vindicated – modernity has organized violence with ever greater efficiency. (Cudworth 2015, p. 15)

Cudworth is describing a socially and culturally contingent set of material entanglements, ways of relating and 'becoming with' that are brutally diametric. Yet 'we'—in the relatively affluent consumer societies of the global North—are routinely shielded from the phenomenal reality of these entanglements, through narrative frames that normalize meat eating in various ways, legitimize nonhuman animals as entertainment, and obscure or belittle the implications for animal suffering (Worthy 2008).

From the Fantastic to the Commonsensical?

While Cudworth limits her recommendations to a call for a more inclusive approach within sociologies of violence, trans-species psychologists set their sights on hopes for social transformation. Bradshaw and Watkins argue that the terminology of liberation psychology, normally applied to human suffering in the context of colonialism, patriarchy and capitalism, should be utilized to highlight and challenge the 'oppression, marginalization, exploitation, forced migration, and genocide that animal communities experience' (2006, p. 8). Elsewhere Bradshaw cites Ignacio Martín-Baró's claims about liberation psychology: 'If we want psychology to make a significant contribution to the history of our peoples … we have to redesign our theoretical and practical tools, but to redesign them from the standpoint of the lives of our own people: from their sufferings, their aspirations, and their struggles'. She continues, 'the same holds when considering other species' (2010, p. 416).

Beyond critique, the constructive emphasis of a trans-species approach is the development of an alternative paradigm which encourages empathy, affinity and connectivity; 'a shared trans-species being-in-the-world constituted by complex relations of trust, respect, dependence, and communication' (Wolfe 2010, p. 141); in other words, an alternative narrative framing of a set of practices that acknowledge 'entanglement' as an ethical relationship and responsibility.

The aim here is not merely to encourage more positive attitudes towards nonhuman animals but to unsettle and expose taken-for-granted understandings of our relations with other species. For Bradshaw and Watkins, this can only be achieved via 'the de-privileging of human language and a renewed reconnection with all other beings' forged out of 'new forms of listening, alternative modalities of communication' (2010, p. 14). This, they claim, is the basis for 'the creation of a new episteme (a new way of seeking knowledge) and a new praxis' (p. 13). For the authors, the act of de-privileging is a necessary first step in 'deconstructing psychologi-cal and cultural privilege' at the 'radical edge of co-rights of animals and humans' (p. 14).[2] They offer trans-species psychology as:

> a way to openly acknowledge the paralyzing grief and trauma in environ-mental destruction... In seeking to deconstruct animal oppression, we are charged at the same time to engage in deep introspection and perhaps the reinvention of our own ontologies. To do so opens the possibility for regen-erating ecological cultures and a way to call back the animals. (Bradshaw and Watkins 2006, p. 21)

Articulating the precise nature of a 'renewed reconnection', 'new forms of listening' or 'alternative modalities of communication' is more difficult. However, there are attempts to address 'material and emotional connections to nonhuman species' (Potts 2010) more concretely, within and beyond trans-species psychology. If we wish to flesh out the contours of an alter-native narrative based on a different understanding of human-nonhuman 'entanglements', it is necessary at this juncture to explore them a little further.

Becoming with

Trans-species psychology includes a focus on human-animal interac-tions, with efforts made to 'place value on more-than-human animals as genuine dialogic participants in the world' Schutten 2015, p. 180); and to decipher, however uncertainly, both sides of this dialogue (Bradshaw 2009; Garcia 2014). Similar research has recently been undertaken in anthropology (e.g. Dugnoille 2014; Faier and Rofel 2014; Kohn 2013; Tsing 2012), communication studies (Plec 2013); cultural studies (Herman 2013; Willett 2013); and beyond (Bankoff 2014; Matsuoka and Sorenson 2014). However, translating such an emphasis into research poses various methodological and theoretical challenges, even for inter-disciplinary fields avowedly dedicated to its study (Birke and Hockenhull 2012; DeMello 2012). This is partly because it is breaking new ground; but also perhaps because it is an attempt to articulate a process that inher-ently exceeds human forms of language and interaction.[3]

One particular interspecies relationship that has been approached from various theoretical and methodological perspectives in the context of anthropogenic ecological degradation is the encounter between human beings and birds. It is a relationship which has often been the focus of environmentalist thinking, evident in one of the founding texts of the contemporary environmentalist movement:

> There was a strange stillness. The birds, for example – where had they gone? Many people spoke of them, puzzled and disturbed. The feeding stations in the backyards were deserted. The few birds seen anywhere were moribund; they trembled violently and could not fly. It was a spring without voices... only silence lay over the fields and woods and marsh (Carson 1962 [1971], p. 22)

It is also the scene, from which Rachel Carson draws the book's title, *Silent Spring*. Birds have long been perceived as markers or omens of the state of the relationship between the human and more-than-human world (Wormworth and Sekercioglu 2011). The Birdlife International and National Audubon Society synthesis of global data on bird populations in relation to climate change (2015), the largest report of its kind at the time of publication, makes this role explicit, in the title—*The Messengers*—and subsequent introduction:

> Over time and across cultures, birds have sent us signals about the health of our environment. The canary in the coal mine offered that most precious resource, time – a small window in which humans could escape toxic gases. Miners no longer use song-birds as early warning systems, but birds are our closest connection to wildlife on the planet and they still tell us about the health of the places people and birds share. Never before has their message – climate change is here and a threat to the survival of birds and people – been as clear or as urgent. (2015, p. 3)[4]

The loss of bird life is also claimed by some to be the basis for developing narratives that can galvanize collective action in response to anthropogenic ecological degradation. Dickinson's application of Terror Management Theory, discussed at length in Chap. 6, attempts to develop this claim. According to Dickinson, Becker looks to the past to claim that the natural world 'was [once] an integral part of immortality-striving rituals and symbolism, providing a context for the vital lies or "character armour" that people require to survive as conscious beings in a social world' (Dickinson 2009, p. 42).

Dickinson sees the potential for the incorporation of the nonhuman natural world in the *present*, however, where other species become the 'ideological symbols or charismatic archetypes' that form the basis for immortality striving. In the context of significant increases in membership of ornithological societies in the USA, she speculates optimistically about the potential of birds and flight as totems for alternative 'immortality-striving hero systems': 'Like fantasy and belief in the supernatural, the idealization of birds may have anxiety-buffering effects' (2009, p. 43 see also Cohen et al. 2011).[5] Dickinson derives from Terror Management Theory a call for the incorporation of idealized nonhuman others into an overarching narrative. Her hope is that doing so might serve the twin purpose of creating a cultural formation that buffers death-related anxiety and encourages empathic identification with the nonhuman natural world.

> By merging two ideas, i.e., the personal connection to nature through birds … and the denial of death… the growing attachment to and projection-idealization of birds and other charismatic species may provide an important route to helping the public to recognize, care about, and act upon climate change in a sustained way. (Dickinson 2009, p. 44)

Although Dickinson applies the logic of Terror Management Theory with conviction, there is an absence here—the value of *experiences* of reciprocity with nonhuman nature upon which alternative narratives might be built—the 'embodied entanglements' described earlier. While Dickinson points to the importance of a 'personal connection to nature' through birds, her argument is largely cast in terms of *thinking with* rather than *being with* birds. She does not consider whether there is a need for the 'new forms of listening' or 'alternative modalities of communication', dear to trans-species psychology (Bradshaw and Watkins 2006, p. 14).

The embodied entanglements of human-bird are not especially relevant because, in Dickinson's account, birds are standing in for something; they are 'merely' totemic, and still primarily significant in terms of human psychological projections and idealizations. However, a primary tenet of trans-species psychology is that 'the construct of psyche decoupled from other species precludes interspecies' relationships with animals in ways *other than* as object or projection' (Bradshaw and Watkins 2006, p. 2; emphasis added). This suggests a broader point—that for nonhuman nature to 'matter' to us, the social and cultural framings that have been described here as fundamental to our embodied sense of meaning and engagement with the world—to our ontological security—have to tangibly incorporate that reciprocity *as* meaningful. Eduardo Kohn advances an argument that builds on similar logic, but from a very different starting point—his anthropological work with the Runa people of Ecuador's Upper Amazon (2013).

Capacious Relationality

Kohn's book recounts his attempt to develop a deeper understanding of the relationship between human and more-than-human worlds, based on his ethnographic fieldwork among the Runa people. The perspective he develops is built on a complex account of the nature of representation, but along the way he offers tantalizing glimpses of what he refers to as 'capacious relationality'. The clearest way to convey this concept is via one of Kohn's own experiences, which he recounts in the book. He recalls a bus journey through Ecuador's Amazonian region east of the Andes. Kohn shares the bus with a number of locals and tourists, as it makes its way via a winding route to the region of Oriente. The journey is abruptly halted by a series of landslides, which have been triggered by heavy rain. As a result, rocks completely block the road ahead and behind. While attempts are made to clear the rocks, the bus is stationary, and, Kohn feels, vulnerable to further landslides: 'The mountain above was starting to fall on us. At one point a rock crashed down onto our roof. I was scared' (2013, p. 46).

Kohn is shaken and unsettled by the experience. As they wait, his feeling of unease is significantly exacerbated by the apparent indifference to the situation displayed by his fellow passengers. Displays of nonchalance from the driver and locals he partially understands—they may well have had similar experiences before. It is the tourists' reaction that especially unnerve him:

> As I worried, these women were joking and laughing. At one point one even got off the bus and walked ahead a few cars to a supply truck off of which she bought ham and bread and proceeded to make sandwiches for her group. The incongruity between the tourists' nonchalance and my sense of danger provoked in me a strange feeling... what at first began as a sense of unease soon morphed into a sense of profound alienation... Because I sensed that my thoughts were out of joint with those around me, I soon began to doubt their connection to what I had always trusted to be there for me: my own living body, the body that would otherwise give a home to my thoughts and locate this home in a world whose palpable reality I shared with others. I came, in other words, to feel a tenuous sense of existence without location – a sense of deracination that put into question by very being. (2013, p. 47)

The feeling of unease stayed with Kohn for the rest of the day, and of interest here is what he has to say a little later about how his disquiet is resolved:

> The next morning after a fitful night I was still out of sorts. I couldn't stop imagining different dangerous scenarios, and I still felt cut off from my body and from those around me. Of course I pretended I wasn't feeling any of this. Trying at least to act normal, and in the process compounding my private anxiety by failing to give it a social existence, I took my cousin for a short walk along the banks of the Misahuallí River... Within a few minutes I spotted a tanager [a small to medium-sized species of bird] feeding in the shrubs at the scruffy edges of town where molding cinder block meet polished river cobbles. I had brought along my binoculars and managed, after some searching, to locate the bird. I rolled the focusing knob and the moment that bird's thick black beak became sharp I experienced a sudden shift. My sense of separation simply dissolved. And, like the tanager coming into focus, I snapped back into the world of life. (2013, p. 47–8)

Kohn makes an interesting association between his experience and a constructionist account of 'panic attacks' that tracks an individual's (Meg) experience of them (Capps and Ochs 1995). Kohn attributes his disquiet to his own, and Meg's, fear being *experienced* as legitimate, but then not being socially *recognized* as such. Following Bion, this disconnect means the subjective anxiety cannot be meaningfully *contained* socially—there is nowhere for it to 'go'. The consequence is that his 'symbolic thought' runs away with itself, creating a radical separation from 'the indexical grounding their bodies might otherwise provide' (2013, p. 49); indexical ties with others (human and nonhuman) ground thinking in our bodies and the world around us. Kohn's experience of profound disjunction leads to a need to 'reground' if possible, to be able to locate oneself again within a web of meanings.

In Kohn's example, successful regrounding does not depend on rarefied mental representations or rationally organized 'cultures of honour' in the way Dickinson implies; they are spun in relationship, between the material nodes of 'others'. Humans can provide these fleshy reference points, but Kohn's basic point, amongst others, is that there is inherent value in expanded fields of reciprocity that embed (or contain) us in our wider nature; and also that this wider nature is much more capable of doing so than a contemporary anthropocentric viewpoint would suggest. It is in relation with the nonhuman natural world that we might come to 'appreciate that some things that matter can only be fully experienced when we are not entirely reliant on our own symbolic representation of them' (Henderson 2012, p. 181). Although wary of romanticizing any particular experience in nature, Kohn claims that 'sighting that tanager in the bush at the edge of town taught me something about how immersion in this particularly dense ecology amplifies and makes visible a larger semiotic field in which we are all—usually—emplaced' (2013, p. 49). It is this appreciation that provides us with a meaningful embeddedness in the world that human symbolism alone cannot achieve; akin to discovering an unwritten, unspoken language that we share with other creatures and forms of life.[6]

Kohn's approach is useful because it does not simplistically assert the value of a direct relationship between psyche and nature, in which culture is perceived as a mere obstacle; but it does not reduce meaningful experience to human language and what can be narrated. The cultural work of meaning creation and narrative frames are still central, but in valorizing reciprocity with nature, there is the potential to amplify and make 'visible a larger semiotic field beyond that which is exceptionally human' as noted above. Feelings of 'radical separation', what we might refer to as ontological insecurity perhaps are overcome by being able to contain it 'within something broader'.

In cultures where reciprocity with nature is largely written out of the worldview, this 'capacious relationality' is harder to establish. Kidner argues that the opacity of attempts to articulate alternative frameworks is to be expected—they must emerge from a situation where such understandings are marginalized and 'ideologically occluded' (Kidner 2001). Nonetheless, Kohn suggests it is possible to radically reformulate the way we approach human-nonhuman relationships, and to experience and articulate relationships between species in new ways.

An Occasional Touch of Otherness

Continuing with the theme of human-bird encounters, Bernacchi's study of bird-watching communities (2013) is fascinating in this context. Her research offers an account of an activity in which birds are symbolic figures at the heart of an elaborate ritual. Her work, therefore, lends itself to a reading that is sympathetic to Dickinson's account. However, the detail of her analysis provides a fuller appreciation of how the power of rituals depends upon moments of embodied engagement with the more-than-human world that has parallels with Kohn's 'capacious relationality'. It, therefore, offers a glimpse of the possible interplay of embodied entanglements, social practices and narrative frames.

Bernacchi draws on theories of ritual communication (such as Rappaport 1979; Douglas 2003) to understand bird watching as a ritual—involving initiation, rites and performances. Up to a point, her description of bird watching parallels how social practice theory conceptualizes 'hobby' activities, such as floorball, photography, or Nordic walking (Pantzar and Shove 2010; Pantzar et al. 2005). As this book has documented at some length, however, social practice theory has tended to neglect the complexity of subjective, affective and interpersonal dynamics of engagement in social practices.

Bernacchi's account of bird watching is interesting in part because her understanding of ritual, informed as it is by a different theoretical tradition, has a different quality to the descriptions of hobbies we find in social practice literature. In successfully 'performing' the practice of bird watching, the meaning of embodied engagement is made absolutely central. The practice hinges on 'recodify[ing] the value of the ordinary object [the bird] into a sacred subject' (2013, p. 146). To elaborate briefly, though rituals imply formality and structure, liminal moments are essential to their 'sacred' dimension (Curtin 2009). These are instances of uncertainty or inexplicability, experienced as spontaneous and embodied, individual but contributing to a collective, and, essentially, *prior* to being identified within an agreed taxonomy. With respect to 'birding' (the preferred terms of Bernacchi's participants), the liminal moment is the instance upon which the whole enterprise hinges: 'the manifestation of birds' (2013, p. 147):

> Once a participant chooses to engage in the ritual and has learned some of the ritual of birding's structures, there is an in-between place experienced when the bird arrives and none of the birders have identified it. This is a rite of transition, where life is different upon seeing and identifying the bird... The space through which the birders passed is known as the liminal space, a threshold where everyone is relinquished from their roles and still engaged. (2013, p. 147)

What is significant about Bernacchi's reading of birding as ritual here is the sense in which the moment of observation is an embodied experience of encounter with non-narrative reality. Despite the structure, formality and symbolism of birding, it still 'depends on what happens *before* our animal eyes shift from the more-than-human world' (2013, p. 144). Birder's encounters with birds are often brief, enabling a momentary suspension of normal awareness: 'to be aware of a (fleeting wildlife) experience means that it has already passed' (Norretranders, cited in Curtin 2009, p. 465). This liminal moment is prior to any symbolization, but essential to the 'object' becoming sacred within an order of signs. The liminal moment is akin to Kohn's recognition of the tanager as offering a pre-symbolic grounding of the human self in wider nonhuman ontology—a capacious relationality.

There is a danger here of course that we forget how much we reside on the 'anthropocentric side' of ritual communication between humans and birds (Bernacchi 2013, p. 144). Furthermore, as an example, bird watching might appear to be stretching the definition of the *encounter*, suggesting something more akin to an act of spectatorship. Curtin's description of human-nonhuman animal contact certainly places greater emphasis on the importance of mutuality: 'An 'encounter' with a wild animal suggests more of a meeting, of being in a shared space with both animal and human being acutely aware of one another especially where there is one-to-one experience or where there is eye contact' (2009, p. 466).[7] In this regard, Bernacchi finds solace in Peters' modest assertion that human and more-than-human connection might best be sought fleetingly, rather than in a grand melding of consciousness: 'If we thought of communication as the occasional touch of otherness rather than a conjunction of consciousness, we might be less restrictive in our quest for nonearthly intelligence' (cited in Bernacchi 2013, p. 144).

Encountering Ravens

Munday more explicitly grapples with the possibility of *mutual* recognition between humans and birds in her account of human-raven interaction. Munday avowedly attempts to articulate a human connection with ravens *beyond* one which posits the latter as 'merely' totems or human projections, a constituent of formulaic myths and conventions: 'Instead we can interact with wild animals such as ravens, construct meaning on a day-to-day basis as we go and try to understand the world through them' (2013, p. 216).

She offers numerous examples of the complexity of raven interaction and communication with other ravens and with other species, particularly in relation to hunting and foraging tactics.[8] She also describes human-raven interaction, such as the way a raven can dip a wing as a 'performative act'—it apparently acts as a sign to a human hunter, indicating the location of elk or similar prey, with ravens subsequently foraging carcass remains. Munday offers an example drawn from her own experience walking in the northern Rocky Mountains:

I rarely climb a mountain peak without having a flock of ravens show up as I scramble along the ridge. They play on the thermals and seemingly take a perverse delight in surprising me. Raven joy is as apparent as a playful dog's joy: They approach from behind, call out as they fly above the ridgeline and then quickly drop down as I turn to see them. This is repeated. What are they up to? Do they expect me to kill and share a mountain goat with them? Are they 'pulling my tail,' the way they tease wolves? It remains a mystery, but is fun to think about. It *is* constructive interplay. (2013, p. 217; emphasis in the original)

Munday considers human-animal interaction to be most significant when it has this 'material-semiotic' dimension. For Munday, ravens themselves are authors of stories, communicated to us, even if we do not have direct access to raven thought and understanding. So when a raven indicates to a human or wolf the location of an elk, so that it might scavenge, it 'models a possible world and communicates that story to another species' (2013, p. 219). Her emphasis on embodied communication encourages us to think of animal encounters not just as a basis for alternative narrative forms, but as dialogical partners in the ongoing construction of narratives.

Trans-Species Identifications as a Basis for Sustainable Practices

Munday makes something of a leap in connecting experiences of this nature with the possibility of transformative social change. Engaging with wild nature, she asserts, 'can profoundly change our relationship to the world. From this perspective, we can imagine a new future' (2013, p. 219). In a similarly speculative vein, Bernacchi sees the potential for change in the birder's practical pursuit of wild birds, which takes them into, often, novel interstices of urban, wasteland and human-made sites. This practice, she hopes, can unsettle the 'tenuously perched dichotomy' of what we have come to accept as 'Nature and not-Nature, inviting the meaningful recognition of a more complex ecology' (2013, p. 156).

Evidence of the difference that trans-species identifications make to sustainable living, whether we count this as policy making, conservation, protest or community activism, is harder to establish. However, in feminist philosophy and psychology, object relations theory and developmental and social psychology, there are a number of theoretical approaches that locate recognition of affiliation and interdependence with others as the basis for more ethical and compassionate relationships towards those others (e.g. Gilligan 1982; Belenky et al. 1986). Following the important work of Winnicott and others on the formative importance of relational space for development, human beings are conceptualized as fundamentally inter-subjective and co-emergent. Our experience of self-in-other/other-in-self is unconscious, forged in our prenatal experiences—the matrixial relations of becoming mother and baby in the womb (Ettinger 2006; Neill 2008).

If we accept that the relation dynamics of early childhood are embodied and pre-verbal but foundational, how is their fuller recognition the basis for more compassionate and ethical relations with others? The presumption tends to be that in acknowledging that we are stitched into the fabric of the purposeful subjectivity of others, we are more likely to relax the imperative for egoic mastery and control; this is certainly Butler's logic (Butler 2005). It is in acknowledging our own incompleteness and our reliance on 'the other' to constitute us as individuals that we have the basis for more ethical relations with others (Adams 2010).

The move that trans-species psychology and numerous interdisciplinary developments make is to extend this understanding—echoing Haraway again (2014)—to a 'multi-species becoming with…in relationality all the way down'. At the heart of this movement is the belief that recognizing fundamental trans-species interconnection and interdependence is, to paraphrase Renee Letzman, an 'optimal context' for facilitating the expression of concern, and reparative practices in the context of anthropogenic ecological degradation (Lertzman, 2015). Interrelationship and interdependence of human and nonhuman nature must be embedded in our ways of understanding the world and embodied in our engagement with it.

A Sense of Belonging

Framed slightly differently, a core purpose of trans-species psychology is, in a sense, the discovery and promotion of an expanded sense of self to include the other, extending the perception of similarity and commonality with other species as a basis for equal rights (Opotow 1993; Presser 2013). Social psychological research has indicated that conceptions of 'human uniqueness'—a corollary perhaps, of egoic mastery and control, are utilized as strategies of moral disengagement—justifying animal suffering in farming practices for example (Bilewicz et al. 2011; see also Bastian et al. 2011; Bratanova et al. 2011; Mitchell 2011). Theoretical underpinnings for social psychological research of this nature owe more to a tradition studying the dynamics of group membership and social categorizarion. Cynthia Frantz and Stephan Mayer explicitly address the broader relevance of this field for ecological crisis and sustainable behaviour (Frantz and Mayer 2014). They attempt to utilize the social psychology of belonging to claim that 'the deep motivation that comes from a sense of "we-ness" is one of the few psychological forces strong enough to compete with the prevailing counterforces required to engage in environmentally responsible behaviour' (Frantz and Mayer 2014, pp. 85–86). Broadly speaking, if we share a sense of 'we-ness' with others, we are more likely to care about them, more motivated to cooperate with them, protect and defend them (Vining 2003).

There is, as yet, little empirical work in social psychology that takes up Frantz and Mayer's challenge by exploring human-nonhuman animal *encounters*. A notable exception is Jackie Abell's research with volunteers on conservation projects for endangered animal species (Abell 2013). In line with Frantz and Mayer, Abell asserts the need for psychology to address the 'processes under which humans develop a sense of responsibility for other living things' and 'the decisions humans make in caring for rather than exploiting and consuming animals' (Abell 2013, p. 158). Many of Abell's participants revealed a strong sense of commonality—or 'we-ness'—with the animals they helped. She found that the strength of such feelings were related to a number of themes, including a tendency to 'anthropomorphize' nonhuman animals, perceiving their excitement, appreciation and need for love for example; but also reciprocity, in experiencing a capability to recognize and respond to human needs (e.g. for 'bonding' and companionship) in nonhuman animals (p. 164; see also Lott 1988). In line with Frantz and Mayer's approach, Abell emphasizes how her participant's experiences allowed them to begin to 'dismantle the demarcation between 'human' and 'animal' such that the act of volunteering established common group membership' (p. 163)

The encounters Abell studies are arguably specific manifestations (though not to be especially valorized) of Plumwood's 'embodied entanglements' with which we had begun this chapter; of the 'material-semiotic dimension' of human-nonhuman animal interaction described by Munday (2013); and the trans-species contact that has become the focus of communication scholars, trans-species psychologists, ethnobiologists and many more. Abell's research, along with these other accounts, tell us in various ways something of the importance of the materiality of interspecies interrelationships—the 'being with' as much as the 'thinking with'; and the potential significance of these experiences for furthering a sustainability agenda.

Attending to Other Ways of Being

Kohn argues that challenging taken-for-granted understandings of the human-nonhuman divide is a political project because pointing to the ways humans can 'relate to other kinds of beings can help think possibility and its realisation differently'. This is prefigurative or 'alter-politics' (Hage 2012) as an imaginative act, not, or not only, based not on opposition to or critique of current systems. Alter-politics 'grows from attention to another way of being, one here that involves other kinds of living being' (Kohn 2013, p. 14). A trans-species perspective encourages a reconceptualization not just of the human, but 'of how ethically bound the human is to other forms of life, bound in our shared vulnerability, to other living beings who think and feel, live and die, have needs and desires, and require care just as we do' (Wolfe 2010, p. 140).

Claims that paying more attention to interspecies interdependencies unproblematically translates into a widespread and unstoppable ecocentric paradigm is naïve at best. However, it is heartening that there is a growing appetite for such work. As recently as 2010 Potts stated that 'a critical psychology of human-animal relations – that is, a psychology focused not on human-animal comparisons but rather on the interactions between humans and nonhuman animals – has been slow to take off and remains both rudimentary and very much on the margins' (Potts

2010, p. 294). It *has* taken off across a range of disciplines even if it is routinely marginalized in psychology and the sustainability agenda.

In pursuing some of the ways in which human and more-than-human entanglements are being articulated today, we contribute to a questioning of the prescriptive powers of the 'rationalist paradigm' to obfuscate and overwrite the relationality at the heart of the ecological crisis. This undertaking 'points us toward the necessity of an ethics based not on ability, activity, agency, and empowerment but on a compassion that is rooted in our vulnerability and passivity' (Wolfe 2010, p. 141). More prosaically perhaps, to borrow the words of Lauren Berlant, it encourages us 'to focus on patterns of attachment we hadn't even yet known to notice' (Berlant 2012).

Notes

1. Mainstream psychology largely avoids the study and analysis of embodied experiences of human and more-than-human nature 'entanglements'. It has long been considered a topic 'tainted' by spiritualism or mysticism—neither of which has a place in psychology as a profession or 'science' (Reser 1995; Vining 2003). Social science approaches keener to stress the power of situational and social contexts have also neglected to explore such experiences, hesitancy here reflecting antipathy towards 'essentialist' understandings of 'nature'; and the ascendancy of social constructionism, in which meaning-making too readily begins and ends within the realm of human capacities (Crist 2004).
2. The logic of how understanding gets translated into action here is analogous to social psychological accounts of group identity and empathic behaviour. When one's sense of self is expanded to incorporate others as part of 'us' in human-to-human relationships, more empathic and protective behaviour tends to follow (Aron et al. 1991; Cialdini et al. 1997); the same dynamics are potentially inherent in moving 'beyond assumed species alignments' (p. 14), towards a more 'capacious relationality' (Kohn 2013). Frantz and Mayer goes as far as too assert that the motivational power of 'a sense of "we-ness" is one of the few psychological forces strong enough' to challenge 'prevailing counterforces' (Frantz and Mayer 2014, pp. 85–6).

3. Vining describes human-animal interaction as 'noetic but ineffable': '[it] is knowable but we are unable to describe it in words' (2003, p. 94).

4. Despite variation in the fortunes of different species, bird life in general is in decline or under threat across the planet (BirdLife International 2014; Runge et al. 2015). Across Europe, over a third of common bird species have declined over the last 30 years, with farmland birds faring particularly badly (BirdLife International 2013). One estimate suggests that there are 300 million fewer birds in Europe today than in 1980 (Pan-European Common Bird Monitoring Scheme 2012). In the UK in particular, official bird population surveys report a marked overall falling-off over the last 40 years; including a 55 % fall in farmland birds; 28 % in woodland birds, 24 % in seabirds and 17 % in wetland birds (DEFRA 2014).

5. Examples of birds as *totems*—symbolic place-holders for myth and rituals—have been noted in many cultures throughout history (Munday 2014), and long acknowledged in anthropology (e.g. Goldenweiser 1912; Hogbin 1934). See Cohen et al. (2011) for a related discussion.

6. Kohn is not the only scholar to assert the need for other kinds of practice, relatedness and consciousness. Bernstein's 'borderland consciousness' (Bernstein 2006), and Fisher's 'recollective practices' (2013a) are parallel calls, embedded in Jungian and ecopsychology traditions respectively, rather than Kohn's anthropological roots.

7. Kahn et al. (2010) offer 'recognition by a nonhuman other' as an example of patterns in human-nature interaction that are often lost or hidden in modern societies, further emphasizing the significance of mutual recognition in human-nonhuman encounter.

8. Such as ravens leading wolves to carcasses that they cannot penetrate with their beaks, so that the wolves will open up the carcasses and make them accessible to the ravens. Munday suggests wolves clearly know a raven's intention in this context, will willingly be guided, and often share a carcass.

References

Abell, J. (2013). Volunteering to help conserve endangered species: An identity approach to human–animal relationships. *Journal of Community and Applied Social Psychology, 23*(2), 157–170.

Adams, C. J. (2015). *The sexual politics of meat: A feminist-vegetarian critical theory* (3rd ed.). London: Bloomsbury.

Adams, M. (2010). Losing one's voice: Dialogical psychology and the unspeakable. *Theory & Psychology, 20*(3), 342–361.

Adams, M. (2015). The wider environment. In J. Turner, C. Hewson, K. Mahendron, & P. Stevens (Eds.), *Living psychology: From the everyday to the extraordinary* (pp. 369–412). Milton Keynes: Open University Press.

Aron, A., Aron, E. N., Tudor, M., & Nelson, G. (1991). Close relationships as including other in the self. *Journal of Personality and Social Psychology, 60*, 241–253.

Bankoff, G. (2014). Learning about disasters from animals. In H. Egner, M. Schorch, & M. Voss (Eds.), *Learning and calamities: Practices, interpretations, patterns* (p. 42). London: Routledge.

Bauman, Z. (1989). *Modernity and the Holocaust*. Cambridge: Polity Press.

Belenky, M., Clinchy, B. M., Goldberger, N. R., & Tarule, J. (1986). *Women's ways of knowing: The development of self, mind, and voice*. New York: Basic Books.

Berlant, L. (2012, June 15). Interview in *Rorotoko: Cutting-Edge Intellectual Interviews*. http://rorotoko.com/interview/20120605berlantlaurenoncruelop timism/?page=1. Accessed 21 Dec 2015.

Bernacchi, L. A. (2013). Flocking: Bird-human ritual communication. In E. Plec (Ed.), *Perspectives on human-animal communication: Internatural communication* (pp. 143–161). London: Routledge.

Bernstein, J. S. (2006). *Living in the borderland: The evolution of consciousness and the challenge of healing trauma*. London: Routledge.

Bilewicz, M., Imhoff, R., & Drogosz, M. (2011). The humanity of what we eat: Conceptions of human uniqueness among vegetarians and omnivores. *European Journal of Social Psychology, 41*(2), 201–209.

Bion, W. R. (2013) [1961]. *Experiences in groups: And other papers*. London: Routledge.

BirdLife International. (2013). *Europe-wide monitoring schemes highlight declines in widespread farmland birds*. Presented as part of the BirdLife State of the world's birds website. Available from http://www.birdlife.org/datazone/sowb/casestudy/62. Checked 7 Apr 2016.

BirdLife International. (2014). *Agricultural intensification has caused the decline of many common bird species in Europe*. Presented as part of the BirdLife State of the world's birds website. http://www.birdlife.org/datazone/sowb/casestudy/141. Accessed 18 Dec 2015.

Birdlife International and National Audubon Society. (2015). *The messengers: What birds tell us about threats from climate change and solutions for nature and people*. Cambridge, UK/New York: Birdlife International and National Audubon Society. http://climatechange.birdlife.org. Accessed 18 Dec 2015.

Birke, L., & Hockenhull, J. (Eds.). (2012). *Crossing boundaries: Investigating human-animal relationships* (Vol. 14). New York: Brill.

Bradshaw, G. A. (2009). Transformation through service: Trans-species psychology and its implications for ecotherapy. In L. Buzzell & C. Chalquist (Eds.), *Ecotherapy: Healing with nature in mind* (pp. 157–166). San Francisco: Sierra Club Books.

Bradshaw, G. A., & Watkins, M. (2006). Trans-species psychology: Theory and praxis. *Psyche & Nature, 75,* 69–94.

Bradshaw, G. A. (2010). You see me, but do you hear me? The science and sensibility of trans-species dialogue. *Feminism and Psychology, 20*(3), 407–419.

Bradshaw, G. A., Capaldo, T., Lindner, L., & Grow, G. (2008). Building an inner sanctuary: Complex PTSD in chimpanzees. *Journal of Trauma & Dissociation, 9*(1), 9–34.

Bratanova, B., Loughnan, S., & Bastian, B. (2011). The effect of categorization as food on the perceived moral standing of animals. *Appetite, 57*(1), 193–196.

Buckley, C., & Bradshaw, G. A. (2010). The art of cultural brokerage: Recreating elephant-human relationship and community. *Spring: A Journal of Archetype and Culture, 83,* 35–59.

Butler, J. (2005). *Giving an account of oneself.* New York: Fordham University Press.

Capps, L., & Ochs, E. (1995). *Constructing panic.* Harvard: Harvard University Press.

Carson, R. (1962). *Silent spring.* New York: Houghton Miffin.

Cassidy, R., & Mullin, M. (Eds.). (2007). *Where the wild things are now: Domestication reconsidered.* London: Berg.

Choy, T. K., Faier, L., Hathaway, M. J., Inoue, M., Satsuka, S., & Tsing, A. (2009). A new form of collaboration in cultural anthropology: Matsutake worlds. *American Ethnologist, 36*(2), 380–403.

Cialdini, R., Brown, S., Lewis, B., Luce, C., & Neuberg, S. (1997). Reinterpreting the empathy–altruism relationship: When one into one equals oneness. *Journal of Personality and Social Psychology, 73,* 481–494.

Cohen, F., Sullivan, D., Solomon, S., Greenberg, J., & Ogilvie, D. M. (2011a). Finding everland: Flight fantasies and the desire to transcend mortality. *Journal of Experimental Social Psychology, 47*(1), 88–102.

Collard, A. (1988). *Rape of the wild: Man's violence against animals and the earth.* Indiana: Indiana University Press.

Crist, E. (2004). Against the social construction of nature and wilderness. *Environmental Ethics, 26,* 5–24.

Cudworth, E. (2015). Killing animals: Sociology, species relations and institutionalized violence. *The Sociological Review, 63*(1), 1–18.

Curtin, S. (2009). Wildlife tourism: The intangible, psychological benefits of human–wildlife encounters. *Current Issues in Tourism, 12*(5-6), 451–474.

Defra. (2014). Wild bird populations in the UK, 1970 to 2013. HM Government, UK. https://www.gov.uk/government/statistics/wild-bird-populations-in-the-uk. Accessed 18 Dec 2015.

DeMello, M. (2012). *Animals and society: An introduction to human-animal studies.* Columbia: Columbia University Press.

Dickinson, J. L. (2009). The people paradox: Self-esteem striving, immortality ideologies, and human response to climate change. *Ecology and Society, 14*(1), 34.

Douglas, M. (2003). *Purity and danger: An analysis of concepts of pollution and taboo.* London: Routledge.

Dugnoille, J. (2014). From plate to pet: Promotion of trans-species companionship by Korean animal activists. *Anthropology Today, 30*(6), 3–7.

Ettinger, B. (2006). *The Matrixial Borderspace.* Minneapolis: University of Minnesota Press.

Faier, L., & Rofel, L. (2014). Ethnographies of encounter. *Annual Review of Anthropology, 43*, 363–377.

Fisher, A. (2013a). *Radical ecopsychology* (2nd ed.). London: Routledge.

Fisher, A. (2013b). Ecopsychology at the Crossroads: Contesting the nature of a field. *Ecopsychology, 5*(3), 167–176.

Franklin, A. (1999). *Animals and modern culture: A sociology of human-animal relations in modernity.* Thousand Oaks: Sage.

Frantz, C. M., & Mayer, F. S. (2014). The importance of connection to nature in assessing environmental education programs. *Studies in Educational Evaluation, 41*, 85–89.

Garcia, T. (2014). *Form and object: A treatise on things.* Edinburgh: Edinburgh University Press.

Gilligan, C. (1982). *In a different voice.* Harvard: Harvard University Press.

Goldenweiser, A. A. (1912). The origin of totemism. *American Anthropologist, 14*(4), 600–607.

Hage, G. (2015). *Alter-politics: Critical anthropology and the radical imagination.* Melbourne: Melbourne University Publishing.

Haraway, D. J. (2003). *The companion species manifesto: Dogs, people, and significant otherness.* Chicago: Prickly Paradigm Press.

Haraway, D. J. (2008). *When species meet*. Minnesota: University of Minnesota Press.

Haraway, D. (2014, September 5). Anthropocene, Capitalocene, Chthulucene: Staying with the Trouble. *Antrhopocene: Arts of living with a damaged planet*, AURA: Aarhus University Research on the Anthropocene. https://vimeo.com/97663518

Harrison, R. (1964). *Animal machines: The new factory farming industry*. London: Vincent Stuart.

Henderson, C. (2012). *The Book of barely imagined beings: A 21st century bestiary*. London: Granta.

Herman, D. (2013). Selfhood beyond the species boundary. *Postmodern Culture, 24*(1).

Hird, M. J. (2010). Indifferent globality: Gaia, symbiosis and 'other worldliness'. *Theory, Culture and Society, 27*(2–3), 54–72.

Hogbin, H. I. (1934). Culture change in the Solomon Islands: Report of field work in Guadalcanal and Malaita. *Oceania, 4*(3), 233–267.

Kahn, P. H., Jr., Ruckert, J. H., Severson, R. L., Reichert, A. L., & Fowler, E. (2010). A nature language: An agenda to catalog, save, and recover patterns of human–nature interaction. *Ecopsychology, 2*(2), 59–66.

Kidner, D. W. (2001). *Nature and psyche: Radical environmentalism and the politics of subjectivity*. Albany: SUNY Press.

Kidner, D. (2012). *Nature and experience in the culture of delusion: How industrial society lost touch with reality*. Basingstoke: Palgrave.

Kirksey, S., & Helmreich, S. (2010). The emergence of multispecies ethnography. *Cultural Anthropology, 25*(4), 545–576.

Kohn, E. (2013). *How forests think: Toward an anthropology beyond the Human*. Berkeley: University of California Press.

Latour, B. (1993). *We have never been modern*. Cambridge, MA: Harvard University Press.

Latour, B. (2005). *Reassembling the social*. Oxford: Oxford University Press.

Lott, D. (1988). Feeding wild animals: The urge, the interaction, and the consequences. *Anthrozoos, 1*, 255–257.

Matsuoka, A., & Sorenson, J. (2014). Social justice beyond human beings: Trans-species social justice. In T. Ryan (Ed.) *Animals in social work: Why and how they matter* (pp. 64–79). Basingstoke: Palgrave.

Mitchell, L. (2011). Moral disengagement and support for nonhuman animal farming. *Society and Animals, 19*(1), 38–58.

Moore, J. L. (2013). Speciesism. *Contexts, 12*, 12–13.

Munday, P. (2013). Thinking through ravens: Human hunters, wolf-birds, and embodied communication. In E. Plec (Ed.), *Perspectives on human-animal communication: Internatural communication* (pp. 207–225). London: Routledge.

Neill, C. (2008). Severality: Beyond the compression of the cogito. *Subjectivity, 24*(1), 325–339.

Opotow, S. (1993). Animals and the scope of justice. *Journal of Social Issues, 49*(1), 71–85.

Pan-European Common Bird Monitoring Scheme (PECBMS). (2012). *Population trends of common European breeding birds 2012*. Prague: CSO.

Pantzar, M., & Shove, E. (2010). Understanding innovation in practice: A discussion of the production and re-production of Nordic Walking. *Technology Analysis & Strategic Management, 22*(4), 447–461.

Pantzar, M., Shove, E., & Hand, M. (2005). Innovations in fun: The careers and carriers of digital photography and floorball. In M. Pantzar & E. Shove (Eds.), *Manufacturing leisure*. Helsinki: NCRC.

Paschen, J. A., & Ison, R. (2014). Narrative research in climate change adaptation – Exploring a complementary paradigm for research and governance. *Research Policy, 43*(6), 1083–1092.

Pickering, A. (2005). Asian eels and global warming: A posthumanist perspective on society and the environment. *Ethics and the Environment, 10*(2), 29–43.

Plec, E. (Ed.). (2013). *Perspectives on human-animal communication: Internatural communication*. London: Routledge.

Plumwood, V. (2002). *Environmental culture: The ecological crisis of reason*. London: Routledge.

Potts, A. (2010). Introduction: Combating speciesism in psychology and feminism. *Feminism and Psychology, 20*, 291–301.

Presser, L. (2013). *Why we harm*. New Brunswick: Rutgers University Press.

Rappaport, R. (1979). *Ecology, meaning, and religion*. Berkeley/California: North Atlantic Books.

Reser, J. P. (1995). Whither environmental psychology? The transpersonal ecopsychology crossroads. *Journal of Environmental Psychology, 15*(3), 235–257.

Runge, C. A., Watson, J. E., Butchart, S. H., Hanson, J. O., Possingham, H. P., & Fuller, R. A. (2015). Protected areas and global conservation of migratory birds. *Science, 350*(6265), 1255–1258.

Ryder, R. (1975). *Victims of science: The use of animals in research*. Michigan: Davis-Poynter.

Schutten, J. K. (2015). Book review: Perspectives on human-animal communication: Internatural communication. In *Environmental communication*, *9*(1): 137–142.

Tsing, A. (2012). Unruly edges: Mushrooms as companion species. *Environmental Humanities*, *1*, 141–154.

Vining, J. (2003). The connection to other animals and caring for nature. *Human Ecology Review*, *10*(2), 87–99.

Willett, C. (2013). Water and wing give wonder: Trans-species cosmopolitanism. *PhaenEx*, *8*(2), 185–208.

Wolfe, C. (2010). *What is posthumanism?* Minnesota: University of Minnesota Press.

Wormworth, J., & Sekercioglu, C. H. (2011). *Winged sentinels: Birds and climate change*. Cambridge: Cambridge University Press.

Worthy, K. (2008). Modern institutions, phenomenal dissociations, and destructiveness toward humans and the environment. *Organization & Environment*, *21*, 148–170.

Worthy, K. (2013). *Invisible nature: Healing the destructive divide between people and the environment*. New York: Prometheus Books.

11

Narrative Foreclosed? Towards a Psychosocial Research Agenda

Looking Back

it seems easier to imagine 'the end of the world' than a far more modest change in the mode of production, as if liberal capitalism is the 'real' that will somehow survive even under conditions of a global ecological catastrophe. (Žižek et al. 1999, p. 55)

a properly turned mythology, and its enactment in ritual, will compel sustainability, just as assuredly as it has heretofore impeded it. (Sherry 2013, p. 214)

In this book I have argued that in accounting for human responses to ecological crisis, existing mainstream psychological approaches and interventions are insufficient. I have drawn together numerous developments across and beyond disciplines that hold out promise for a psychosocial approach to human responses to ecological crisis. To recap briefly, the main contours of a scientific understanding of anthropogenic ecological crisis was outlined in the first chapter. In the second, the ways in which sustainability and behaviour change have to date been framed

© The Author(s) 2016   **235**
M. Adams, *Ecological Crisis, Sustainability and the Psychosocial Subject*,
DOI 10.1057/978-1-137-35160-9_11

in mainstream psychology were summarized and critiqued. The limitations highlighted included an underestimation of the nature and scope of change required; the decontextualization and depoliticization of the 'behaviours' contributing to the ecological crisis, and the related marginalization of the power of conflicting interests; and a tendency to construe citizens as largely passive subjects. A cumulative criticism drawing on all of these issues was identified as a persistent neglect of the importance of social context in constructing the nature of the problem, as the basis for interventions, and in the framing of appropriate solutions.

While it may still be the case that 'many in the social sciences still consider environmental issues—even those that threaten the very foundation of modern society—marginal to the core of their disciplines' (ISCC/UNESCO 2013, p. 48), there is a growing body of innovative interdisciplinary work, drawing on the social sciences and humanities, that does address such issues. In Chapter 4 a case for understanding the ecological crisis and sustainability from a social practice approach was established in detail. The contemporary reworking of social practice theory to make sense of sustainability addresses many of the criticisms levelled at psychological understandings and interventions. As the basic unit of analysis, a social practice is to be found not 'in' the individual who is carrying out a practice, but in the bodily, socially, materially and culturally distributed 'elements and qualities' that make a practice recognizable as a practice. A social practice approach ambitiously incorporates psychological, social and material elements into an explanation of what constitutes unsustainable social practices, as detailed in Chapter 4. In sum, we can agree with Hargreaves that in 'contrast to conventional, individualistic and rationalist approaches to behaviour change, social practice theory de-centres individuals from analyses, and turns attention instead towards the social and collective organization of practices—broad cultural entities that shape individuals' perceptions, interpretations and actions within the world' (2011b, p. 79).

However, the ascendancy of a social practice approach to sustainability has been partnered by an emerging body of criticism. In revisiting the critical points lodged at mainstream psychological approaches, a discussion of the main tenets of social practice theory was developed in Chapter 5. Questions were claimed to remain over the extent to which social practice theory appreciates the fundamental ways in which human social practices are embedded in nonhuman nature; or the extent to which the meaning of ecological crisis – and appropriate responses—are enmeshed with political interests and power dynamics. The third area of critique was that in social practice approaches to sustainability we find a relative neglect of the ways in which we actively contest, negotiate and make sense of the reality of anthropogenic ecological degradation. Combined, these criticisms highlight that there is still plenty of work to be done in developing a fully psychosocial account of the ecological crisis and sustainability.

In light of these limitations, the second half of the book was a search for adjunct and alternative perspectives broadly in line with a psychosocial orientation. Chapter 6 explored how an existential framework might be utilized to understand how the reality of ecological degradation, and human responsibility for it, is negotiated and contested. Attention shifted to a Terror Management perspective, inspired by the philosophy of Ernest Becker, broadened in this chapter to incorporate the concept of ontological (in)security. Although by no means a fully formed perspective on ecological crisis, Terror Management Theory's emphasis on mortality salience as it intersects with cultural frameworks raised interesting questions about the nature of psychosocial engagement with ecological crisis. An understanding of the affective and unconscious dynamics involved was pursued further in Chapter 7. This chapter outlined a psychoanalytic orientation to defence mechanisms as ways of 'knowing and not knowing' about ecological crisis, hampering action and maintaining destructive practices. Chapter 8 returned to Terror Management Theory, alongside social psychoanalysis and sociological approaches to emotion, to establish a more thoroughgoing psychosocial account of defence mechanisms.

Chapter 9 was an attempt to extend this explanation by investigating a growing body of important work emphasizing the constitutive role of narrative in establishing and maintaining defence mechanisms. A focus on narrative inevitably brings us to questions of the boundaries between narrative and non-narrative reality; and thornier questions about how finding a way through anthropogenic ecological crisis rests on experiences that are at best difficult to articulate at the level of psyche and society. These considerations were the basis for Chapter 10, an inquiry into the transformative potential of forms of trans-species identification that contribute to, but also exceed, narrative framings of human and nonhuman nature relatedness.

Narrative Foreclosure: Towards a Psychosocial Research Agenda

The primary contribution that I have intended to make with this book is modest—to better understand the complexity of confronting anthropogenic ecological crisis. A second, related hope is to further challenge the prevailing assumption of 'weak' versions of sustainability, that take societal structures and their subjective dimensions as given whilst ostensibly addressing that crisis (Räthzel and Uzzell 2009). Beyond these intentions, it is no easy task to draw the strands of critical and alternative perspectives together as a foundation for a psychosocial research agenda to address ecological crisis and sustainability issues, aimed at interventions that will incrementally usher in more sustainable behaviours, practices and societies. Nonetheless, however difficult and uncertain, a third and final hope for this book is that it offers a theoretical resource for the development of a critical social science research agenda in this field. So where do we begin? As is clear from the summary above, a recurring theme in this book has been the significance of narrative for a psychosocial perspective, when engaged with critically. In what remains of this chapter, I tentatively set out how this engagement might inform one element of a research agenda, building on the concept of *narrative foreclosure*.

The term narrative foreclosure was originally coined by Freeman (2000, 2011), and refers to 'the conviction that no new interpretations of one's past nor new commitments and experiences in one's future are possible that can substantially change one's life-story' (Bohlmeijer et al. 2011, p. 346). Although developed in a very different context, I am struck by the prescience of this concept as a proxy for the impasse of existing narrative framings of climate change and sustainable development acknowledged by many in social sciences and the humanities (Luke 2015; Randall 2009); but also as a concept that might inform empirical work in this area.

Bohlmeijer and colleagues at the University of Twente's Dutch Lifestory Lab research the connections between personal narratives, mental health and wellbeing.[1] It is in exploring these links that they have developed the concept of narrative foreclosure in recent years (e.g. Bohlmeijer et al. 2014; Westerhof and Bohlmeijer 2012). For these authors narrative foreclosure relates to one's past as well as future, and they define it accordingly as 'the conviction that no new interpretations of one's past nor new commitments and experiences in one's future are possible that can substantially change one's life-story' (Bohlmeijer et al. 2011, p. 346).

Their work aims to develop narrative foreclosure as a 'sensitizing concept' for making sense of mental health in the specific context of *ageing*. The authors build on narrative approaches to identity, following Erikson (1959), McAdams (1993, 2006) and others, in which the construction of a meaningful narrative is central, and use the idea of foreclosure to explain how 'narrative identity development falters in later life'. Findings to date cautiously suggest that high levels of narrative foreclosure in relation to one's past and future may indeed be negatively related to mental and emotional health and wellbeing as we reach older age (Bohlmeijer et al. 2014).

The concept has been applied in the field of social gerontology, health and ageing studies (Griffin and Phoenix 2014; Randall 2013; Smith and Sparkes 2005). The relevance of narrative foreclosure to health is clearest perhaps in relation to chronic and/or highly debilitating health conditions (Smith and Sparkes 2005). In their study of men with severe spinal cord injury, Smith and Sparkes describe harrowing instances of narrative foreclosure in their participants' talk, for example:

Interviewer: I think you still have a lot offer, to your children, society, yourself.
Jamie: I don't think so. No. I'm useless. Nothing. My condition won't improve. No point anymore. I'm no one now. It's a matter of sitting here alone until I die. Life ended for me the day I broke the neck.... I was more a 'doer' before the accident. Now I'm not much. Not much. I'm half my weight now. Believe it or not, I was big and strong. Now I am nothing. Life moves on, without me. That is how it is. How it will always

be. I just survive. No ambitions. Nothing.... Sometimes I don't think I can go on. I do. But life won't improve. It can only get worse...There is no hope in my life.

Bohlmeijer et al. (2011) describe how narrative foreclosure is contingent upon particular constellations of personal, interpersonal, social and societal dynamics (see also Antelius 2007). Relatedly, foreclosure is always 'co-authored'—'drawn together in a mutually dependent system' rather than being the product of individual interpretation (2011, p. 369). Other people can delimit or dominate narrative development, effectively foreclosing on that development. This depends on interpersonal and social dynamics of 'counterproductive' communication such as the activation of stereotypes in discourse. Following a broadly Foucauldian perspective, Bohlmeijer et al. argue that discourses unfold in the context of social structures and ideologies; they 'provide the frameworks in which people order the world and from which they derive their subjectivity. Discourses have their unique array of resources and restrictions, of templates and practices, or the self-storying that people do – or are pressured to do' (p. 368).

For Bohlmeijer et al. discourses of old age are reified in social policies, legislations and institutions which contribute to experiences of narrative foreclosure:

> The dominant discourse of aging as decline in combination with old age as a separate life phase that lacks meaningful roles or even brings structured dependency might contribute to narrative foreclosure: aging might be experienced not in terms of actively growing old but of passively getting old – merely living out the story with a lack of meaningful commitments that has been imposed on older persons in general Bohlmeijer et al. (2011, p. 368)

Here Bohlmeijer et al. clearly reflect Freeman's original understanding of narrative foreclosure as a psychosocial phenomenon (Freeman 2000). The concept can readily be thought of as psychosocial in fact, as it connects up the cultural production and reification of narratives with their expression, 'internalization' and embodiment (Freeman 2000, p. 83). Griffin and Phoenix recognize this psychosocial dimension in defining narrative foreclosure more broadly as relating to 'the degree to which the culture in which one lives fails to provide adequate narrative resources for living one's life meaningfully and productively' (2014, p. 396).

When Will the Story End? Narrative Foreclosure and Ecological Crisis

Increasingly we live in a world where nothing makes sense. Events come and go like waves of a fever leaving us confused and uncertain. Those in power tell stories to help us make sense of the complexity of reality. But those stories are increasingly unconvincing and hollow… those stories have stopped making sense. (Adam Curtis 2015 *Bitter Lake*)

Mark Freeman defines narrative foreclosure as the 'unshakable conviction that it is simply too late to live meaningfully' (2000; p. 83). Described emphatically as it is here, the concept resonates with critical accounts of the stark failure of predominant cultural narratives (endless growth, neo-liberalism, consumerism, frontierism) to meaningfully engage with the increasingly perturbing reality of ecological crisis, and how this registers affectively: echoing the experiences described by Bohlmeijer et al.—the existential despair and lack of hope and self-efficacy that accompanies becoming 'de-storied' (2011, p. 368). Klein describes the situation as one where we 'find ourselves trapped in a kind of narrative loop'; frenziedly trying to 'tell and retell the same tired stories' (2011, p. 23).

Future scenarios are also trapped in this loop, delimitated and dominated by those narratives; by powerful vested interests; but also by collectively dispersed experiences of inertia, fear and despondency. These are the tacit components of narrative frames, echoed in Cohen's description of collective denial, cited earlier and repeated here:

Whole societies may slip into collective modes of denial not dependent on a fully-fledged Stalinist or Orwellian form of thought control. Without being told what to think about (or what not to think about) and without being 'punished' for knowing the wrong things, societies arrive at unwritten agreements about what can be publicly remembered and acknowledged. (2001, pp. 10–11)

These 'unwritten agreements' about what can be 'remembered and acknowledged' apply equally to visions of the future; and can be understood as elements of a collective experience of narrative foreclosure. Sociocultural forms of denial combine with personal, internal dynamics and discourses, 'official' forms—'public, collective, highly organized', often resourced, spread and maintained by the state—as well as 'microcultures' of denial within a range of institutions and organizations, from families to bureaucracies: 'the group censors itself, learns to keep silent about matters whose open discussion would threaten its self-image... organizations depend on forms of concerted ignorance, different levels of the system keeping themselves uninformed about what is happening elsewhere' (Cohen 2001, p. 11).

Escape Routes

A recurring theme in existing work on narrative foreclosure, though often implicit, is the search for ways of *escaping* narrative foreclosure. Bohlmeijer et al., for example, claim the concept could inform therapy and counselling practice (2011); Griffin and Phoenix as a way of basis for understanding, and subsequently challenging, normative gender and age roles (2014). Other narrative scholars more explicitly attend to the issue of how to *disrupt* narrative foreclosure and open up alternative possible futures. The challenge here is to recognize the ways narratives are interrelated with context, including 'non-linguistic elements of signification, the bodies of storytellers and audiences, the physical environment, historical memory, economic determinants, and political contests' (Squire 2012, p. 68); while exploring how a 'capacity to tell new stories' (Griffin and Phoenix 2014, p. 401) can emerge and be demonstrated.

In their case study approach to aging, women and physical activity Griffin and Phoenix describe how an individual's life story can be told, retold and rewritten in expansive and novel forms, trying on different stories that speak to the possibility of different identities, *if* there is a 'widening [of] available narrative resources' (p. 401). Although developed in a very different context, as critically minded researchers, alert to the obdurate nature of existing social and material arrangements to promote narrative foreclosure, their observations of the steps involved in 'widening one's narrative resources' are salutary (pp. 400–402).

The first step involves knowledge—learning the content that makes up different stories; so this content, as a resource, must be sought out or become available somehow. The second step is learning in a different sense—the ability to tell a different story, and the courage to tell it. This is always ongoing, and, vitally, 'highly dependent on context and audience' to successfully be told (p. 400). The opportunity to tell a story and for it to be heard is imperative, as this is essential in stories being experienced as real: 'only upon articulating their stories did many participants recognise the the narratives that they were drawing from and on' (p. 401). The third step is a critical awareness of the origin, content and form of available narratives, which Griffin and Phoenix describe as 'narrative literacy'. This stage is vital, because it encourages us to 'become aware of both the stories [we] tell and how they are connected to the stories that [we] are surrounded by' (p. 401). If we consider these steps a circular and ongoing process, we can add that this critical examination is the basis for stimulating imagination that encourages in us the capacity to try new stories, reinvigorating the first two steps.[2]

These dynamics elide with what Squire has to say about how narratives can produce important breaks from normative ways of existing, especially as we engage with the future:

> Through the possibilities of movement towards the future, in the sense of an opening of a new context, [narratives] register the particularity of difference, dissidence, and the hard-to-understand. Such narrative appeals from elsewhere, and from others, are not merely disruptive or fragmenting; they can be understood as moral appeals of the future. (Squire 2012, p. 67)

The steps described by Griffin and Phoenix echo here, though the power to stimulate imagination is found as much in the gaps between narratives as it is in the narratives themselves. Moments of uncertainty and incoherence are the 'gifts' of narrative that 'allow movement, opening up new contexts and futures, new possibilities for how one might and should live' (Squire 2012, p. 81).

Beyond the Climate Change Imaginary

The concept of narrative foreclosure is important in the context of anthropogenic ecological crisis because it provides a vocabulary for extending our thinking about the problems with existing narrative framings of that crisis, and how they interpenetrate psychological and social contexts. Although they do not use the term explicitly, a number of commentators and researchers, coming from a range of perspectives, understand current framings of climate change in ways that closely parallel the concept of narrative foreclosure, especially as they overlap and intersect with related discourses of consumerism, neoliberalism and individualism (e.g. Randall 2009; Luke 2015). In his recent analysis of the 'climate change imaginary' Timothy Luke asserts the following: 'Combating climate change, it would appear, is just not an effective mobilizing mega-narrative, and the feeble stand-in used by far too many, of 'sustainability,' means at best 'more of the same only in green' or at worst nothing' (Luke 2015, p. 12).

Important for Luke is what is *closed down* by this particular imaginary: 'Other less visible changes, like the prospects for massive methane releases, huge increases in solar radiation levels, constant high heat, super-destructive weather, greater marine acidification or massive biota extinctions rarely, or never, get into the script of the 3-D movie now in production daily' (Luke 2015, p. 291). Randall similarly points to the 'parallel' narratives of climate change as a problem and its solutions as *inhibiting* imaginative and creative responses. As a 'problem' it is narrated as an apocalyptic but far off disaster; as a solution the story is one of 'small step' behavioural change or faith in techno-fixes on the horizon. Although such narrative frames are not the sole vehicles for collective inaction, denial and displacement, their predominance forecloses the development of alternative possibilities. Yet, following the narrative analysis of scholars such as Squire, this foreclosure is never complete, and the possibility for ruptures is immanent in the dynamics of narrative production (Squire 2012).

Beyond critique then, a productive translation of the concept of narrative foreclosure into ecology and sustainability issues is to engage in research whereby narrative foreclosure can be *challenged*. Following the logic of Griffin and Phoenix, to do so is to research contexts in which a 'widening' of available narrative resources, and the opportunity to engage in the steps described above, becomes possible. Work, which aims to foster imaginative capacities in supportive contexts might provide an avenue for challenging narrative foreclosure in the context of anthropogenic ecological degradation. Paschen and Ison are amongst those scholars calling for an emphasis on narrative inquiry in this area, recognizing that 'how we 'story' the environment determines how we understand and practice adaptation, how risks are defined, who is authorized as actors in the change debate, and the range of policy options considered' (2014, p. 1083). In appreciating the value of narrative research, in particular, we are reminded of the centrality of narrative to the creation of meaning, a core theme in this book. Paschen and Ison describe 'narrative research' in this context as follows:

> How a community 'stories' its past experiences and actions ultimately determines how it understands and practices future adaptation. The narrative researcher elicits such stories, using a conversational interview technique with open-ended questions. In this way, he or she learns about local cultures of dealing with flood, but, in listening attentively may also discover how at first seemingly unrelated socio-cultural or institutional aspects ultimately influence local adaptive capacity (2014, p. 1084)

The researcher's role here, along with the participants, is to tease out the connections between ground-up generation of narratives and the rather ambiguously termed, but clearly 'top-down', 'socio-cultural or institutional aspects'—echoing Griffin and Phoenix's description of a critically oriented 'narrative literacy'. In Paschen and Ison's account too, the critical potential of narrative research goes well beyond 'decoding' persuasive frames in the communication of anthropogenic ecological crisis, and adopting and adapting them in the hope of manipulating desirable decision-making and behavioural outcomes.

Such a research programme has the potential to contribute to understanding how 'power *manifests* itself in the communicative situation, including who speaks, who listens, what kind of language is used' (1084; emphasis in original); and to take situated constructions of meaning to be the basis for knowledge and understanding of the issues in hand, rather than 'expert' knowledge—just as Griffin and Phoenix emphasize the importance of 'a critical examination of the stories we tell and why, the stories we want to tell, and – most important – the stories that others are telling, whether we want to hear them or not' (2014, p. 401). A progressive goal of narrative research, for Paschen and Ison, is not simply to categorize the narratives utilized by individuals and communities, to critically deconstruct media, policy and corporate narratives; or to design 'better' forms of communication ('a passive-receptive view of narrative'); but to encourage, and be actively involved in, 'dialogical and participatory knowledge production' (Paschen and Ison 2014, p. 1085).

From this perspective, socially generated narratives are considered central to the building of 'adaptive capacity' in the context of anthropogenic ecological degradation. The authors' stages of narrative research with the 'wider community' significantly parallel Randall and other's approach to supportive communities:

- bringing together diverse knowledge(s)…
- developing story-telling as an inclusive 'everyday language' for communication…
- integrates knowledge-making with decision-making on the ground…
- the surfacing of the various non-climate drivers of adaptive capacity
- [and] the creation of shared understandings of the meanings and purposes of adaptation. (Paschen and Ison 2014, p. 1089)

In this sense, instigating processes of community dialogue via narrative is a productive process in itself. Ideally, as a 'final' stage, this dialogical approach to narrative production eventually leads to 'new narratives that provide a space for people to perform and enhance their identities within a context of change' (Boxelaar et al., cited in Paschen and Ison 2014, p. 1089). The researcher's role becomes one of asking questions that encourage and elicit reflection, dialogue and the elicitation of 'alternative narratives' (2014, p. 1089).

In an important sense Paschen and Ison's research agenda corresponds with many other psychosocial accounts, such as Randall's emphasis on the significance of sympathetic social support if troublesome affective responses are to be expressed and 'contained'; and if alternative narratives are to emerge and take hold.[3] Marshall similarly emphasizes the importance of the social 'proof' of 'peer transmission' and allegiance rather than the scientific proof of 'peer review' for narratives stressing a need for action to attain validity: 'We follow the social cues of the people we know and trust – our friends, families and preferred media – in the selection of our preferred story of climate change. When we have views that conflict with the social norm around us we choose to suppress them rather than endanger our social allegiances' (Marshall 2014b, p. 97). Hope, following Marshall, resides in the sparks that can shift those norms and allegiances; sparks that can only be ignited via the creation of space for open, meaningful, supportive dialogue and support; facilitating 'opportunities to contribute-to, create, be heard, respected and valued' (Lertzman 2015, p. 146).

Paschen and Ison assert that the authentic acknowledgment of anthropogenic ecological degradation demands, following Krippendorf, 'authentic communication'. They quote him at length to make their point: 'the pleasure of participating in togetherness in which one is free to speak for oneself, not in the name of absent others, not under pressure to say things one does not believe in, and not having to hide something for fear of being reprimanded or excluded from further conversation' (Krippendorf, cited in Paschen and Ison 2014, p. 1089). Again there are clear parallels with a social psychoanalytic emphasis on suitable support structures for affective responses to emerge and be 'held'. Narrative approaches more generally similarly assert that the advocacy of ontologically coherent alternative narratives requires validation from others:

> The individual must have a community of listeners able and willing to hear and validate their experiences in order to create more coherent narratives, and when they do, the evolving narrative coherence is linked to higher levels of both physical and psychological well-being... Thus narrative and identity are dialectically related, such that coherent narratives help create a coherent sense of self, and a coherent sense of self helps create and maintain coherent narratives. (Fivush 2010, p. 96)

The concept of narrative foreclosure and related research provides insights into the working of narrative, and its critical potential on a micro-level, that can contribute to more macro-level understandings of the role of narrative frames in fixing and delimiting our responses to anthropogenic ecological crisis; but also the potential for the development of alternatives. Attending to narrative does not provide a quick-fix remedy to the psychosocial dimensions of ecological crisis, but it offers one route into a critical understanding while providing a basis, however modest, for a research agenda motivated by compassion and hope.

The Contours of What is Possible

To what extent can the kind of research agenda sketched out here incorporate the developments in trans-species psychology, posthumanities and related fields, described in the previous chapter? As noted in the previous chapter, exploring human and more-than-human world relationships empirically poses a number of challenges. This is thanks in part to the noetic nature of this interaction; but also because nonhuman 'others' are routinely outside of the range of convenience of epistemological and methodological norms in most academic traditions. Attempts to bring the nonhuman in to research to date, therefore, varies from theoretical accounts and descriptions of animal-human interaction to more thoroughgoing attempts, to incorporate the nonhuman into the research process.

'Multi-species ethnography' is a promising development in this context (e.g. Fuentes 2012; Smart 2014; Tsing 2012). The term refers to a loose affiliation of anthropologists who have begun 'to place a fresh emphasis on the subjectivity and agency of organisms whose lives are entangled with humans' (Kirksey and Helmreich 2010, p. 545). Examples include human and macaque monkey interaction in shared spaces in Bali (Fuentes 2010); human-horse 'co-domestication' in horse riding communities (Maurstad et al. 2013; see also Birke and Hockenhull 2015); and ethnographies of workplaces defined by human-nonhuman animal interaction such as abattoirs, laboratories and farms (Hamilton and Taylor 2012; see also Birke 2003). Some, though by no means all, of this work explicitly strives to provide 'case studies that place value on more-than-human animals as genuine dialogic participants in the world' (Schutten 2015, p. 2); giving form, however uncertain, to Plumwood's 'embodied entanglements'.

However, incomplete, we are here witnessing attempts to explore and understand human-nonhuman encounters and develop narratives that can adequately frame them. Some, but by no means all, of this work explicitly examines the potential of alternative understandings as a basis for a world-view that is more ecologically sustainable. Growing academic consideration reflects a contemporary cultural preoccupation with the complexity of non-human animal experience, and more specifically, with a renewed inquiry into the ethics of the human-nonhuman animal relationship (Richards 2010, p. 241). This attentiveness is evident in forms as diverse as fiction, memoir, documentary, popular science, photography; alongside campaigns and policies defending and extending the rights of nonhuman animals.[4]

Explorations of human and more-than-human world entanglement are attempts to feel a way though some of the conceptual territory I earlier described as opaque and prefigurative; but also the psychosocial terrain which emphasizes the 'often unrecognized, vague and fuzzy spaces inbetween forms of reality, knowledge and practice' (Brown and Stenner 2009, p. 49).

To make such attempts is to continue to hope—'the key organizing principle' of ecology movements (Mason 2014, p. 154). Rebecca Solnit reminds us that radical hope 'is not about what we expect. It is an embrace of the essential unknowability of the world, of the breaks with the present, the surprises' (cited in Mason 2014, p. 154). For the final word, remaining with hope, I will turn to Nancy Hollander: 'our salvation lies in recognizing the importance of and engaging in the act that goes beyond the horizons of what appears to be possible and redefines the very contours of what is possible' (Hollander 2014, p. 29).

Notes

1. See the Lab's website for more details https://www.utwente.nl/igs/lifestorylab/
2. In fact, imagination is considered an overlooked 'core dynamic' by some sociologists and social psychologists in processes of emancipatory social and cultural change (Zittoun and Gillespie 2015; Frank 2010).

3. It also echoes the calls of Sayer and others for policy makers to engage with people as conscious, concerned citizens, rather than developing ethically and politically questionable sustainability 'behaviour change' programmes that are modeled to work 'behind-the-backs' of their target populations (Sayer 2013; see also Soron 2010).

4. Novels include the aforementioned, Booker prize nominated, novel by Karen Joy Fowler, *We Are All Completely Beside Ourselves* (2013); memoirs include Helen Macdonald's *H Is For Hawk* (2014; winner of the Samuel Johnson Prize and the Costa Book of the Year award in 2014); artistic projects include Jo-Anne McArthur's *We Animals* photo-documentation work http://weanimals.org and Chris Jordan's *Midway* series http://www.chrisjordan.com/gallery/midway/#CF000313%2018x24; documentaries include Blackfish (Director: Cowperthwaite 2013) and *Speciesism: The Movie* (Director: Devries 2013); popular science tacking species relationships and complexity include Caspar Henderson's *Book of Barely Imagined Beings* (2013); Whitehead and Rendell's meticulously researched argument that whales and dolphins have a collective culture (Whitehead and Rendell 2014); campaigns focussing on advancing the legal rights of nonhuman animals include the Arcus Foundation http://www.arcusfoundation.org and the Nonhuman Rights Project http://www.nonhumanrightsproject.org. There is some interlinking of these phenomena. Campaigns against the treatment of Orca whales in commercial aquariums, for example (e.g. http://www.seaworldofhurt.com/news/) is inspired in part by the documentary *Blackfish* and has made some impact on policy making and business practices in the US. See http://www.theguardian.com/us-news/2015/nov/09/seaworld-end-orca-whale-shows-san-diego for details.

References

Antelius, E. (2007). The meaning of the present: Hope and foreclosure in narrations about people with severe brain damage. *Medical Anthropology Quarterly, 21*(3), 324–342.

Birke, L. (2003). Who – Or what – Are the rats (and mice) in the laboratory? *Society & Animals, 11*(3), 207–224.

Birke, L., & Hockenhull, J. (2015). Journeys together: Horses and humans in partnership. *Society & Animals, 23*(1), 81–100.

Bohlmeijer, E. T., Westerhof, G. J., Randall, W., Tromp, T., & Kenyon, G. (2011). Narrative foreclosure in later life: Preliminary considerations for a new sensitizing concept. *Journal of Aging Studies, 25*(4), 364–370.

Bohlmeijer, E. T., Westerhof, G. J., & Lamers, S. M. (2014). The development and initial validation of the narrative foreclosure scale. *Aging and Mental Health, 18*(7), 879–888.

Brown, S. D., & Stenner, P. (2009b). *Psychology without foundations: History, philosophy and psychosocial theory*. London: Sage.

Cohen, S. (2001). *States of denial: Knowing about atrocities and suffering*. New York: Wiley.

Curtis, A. (2015). (Director) *A bitter lake*. London: BBC.

Fivush, R. (2010). Speaking silence: The social construction of silence in auto-biographical and cultural narratives. *Memory, 18*(2), 88–98.

Frank, A. W. (2010). *Letting stories breathe: A socio-narratology*. Chicago: University of Chicago Press.

Freeman, M. (2000). When the story's over: Narrative foreclosure and the possibility of self-renewal. In M. Andrews, S. Slater, C. Squire, & A. Treacher (Eds.), *Lines of narrative: Psychosocial perspectives* (pp. 245–250). Toronto: Captus University Publications.

Freeman, M. (2011). Narrative foreclosure in later life. In G. Kenyon, E. T. Bohlmeijer, & W. R. Randall (Eds.), *Storying later life; issues, investigations, and interventions in narrative gerontology* (pp. 3–19). Oxford: Oxford University Press.

Fuentes, A. (2010). Naturalcultural encounters in Bali: Monkeys, temples, tourists, and ethnoprimatology. *Cultural Anthropology, 25*(4), 600–624.

Fuentes, A. (2012). Ethnoprimatology and the anthropology of the human-primate interface. *Annual Review of Anthropology, 41*, 101–117.

Griffin, M., & Phoenix, C. (2014). Learning to run from narrative foreclosure: One woman's story of aging and physical activity. *Journal of Aging and Physical Activity, 22*(3), 393–404.

Hamilton, L., & Taylor, N. (2012). Ethnography in evolution: Adapting to the animal "other" in organizations. *Journal of Organizational Ethnography, 1*(1), 43–51.

Hargreaves, D. (2011a). Pro-environmental interaction: Engaging Goffman on pro-environmental behaviour change. *Centre for Social and Economic Research on the Global Environment Working Papers 11-04*. Norwich: University of East Anglia. http://www.cserge.ac.uk/sites/default/files/2011-04.pdf. Accessed 18 Dec 2015.

Hargreaves, T. (2011b). Practice-ing behaviour change: Applying social practice theory to pro-environmental behaviour change. *Journal of Consumer Culture, 11*(1), 79–99.

Hollander, N. C. (2014). *Uprooted minds: Surviving the politics of terror in the Americas*. London: Routledge.

ISSC/UNESCO. (2013). *Summary: World social science report 2013: Changing global environments*. Paris: OECD Publishing and UNESCO Publishing.

Kirksey, S., & Helmreich, S. (2010). The emergence of multispecies ethnography. *Cultural Anthropology, 25*(4), 545–576.

Klein, N. (2011). On precaution. In P. Kingsnorth & D. Hine (Eds.), *Dark Mountain Issue 2* (pp. 20–25). Dark Mountain Project: Ulverston.

Lertzman, R. (2015). *Environmental melancholia: Psychoanalytic dimensions of engagement*. London: Routledge.

Luke, T. W. (2015). The climate change imaginary. *Current Sociology, 63*(2), 280–296.

Marshall, G. (2014a). *Don't even think about it: Why our brains are wired to ignore climate change*. London: Bloomsbury.

Marshall, G. (2014b). Five. In J. Smith, R. Tyszczuk, & R. Butler (Ed.), *Culture and climate change: Narratives* (Vol. 2, pp. 96–97). Cambridge: Shed.

Mason, K. (2014). Becoming Citizen Green: Prefigurative politics, autonomous geographies, and hoping against hope. *Environmental Politics, 23*(1), 140–158.

Maurstad, A., Davis, D., & Cowles, S. (2013). Co-being and intra-action in horse–human relationships: A multi-species ethnography of be (com)ing human and be (com)ing horse. *Social Anthropology, 21*(3), 322–335.

McAdams, D. P. (1993). *The stories we live by: Personal myths and the making of the self*. New York: Morrow.

McAdams, D. P. (2006). The redemptive self: Generativity and the stories Americans live by. *Research in Human Development, 3*(2–3), 81–100.

Paschen, J. A., & Ison, R. (2014). Narrative research in climate change adaptation – Exploring a complementary paradigm for research and governance. *Research Policy, 43*(6), 1083–1092.

Randall, R. (2009). Loss and climate change: The cost of parallel narratives. *Ecopsychology, 3*, 118–129.

Randall, W. L. (2013). The importance of being ironic: Narrative openness and personal resilience in later life. *The Gerontologist, 53*(1), 9–16.

Richards, G. (2010). Psychological use of animals. In *Putting psychology in its place: Critical historical perspectives* (3rd ed., pp. 233–244). London: Routledge.

Sayer, A. (2013). Power, sustainability and well-being: An outsider's view. In E. Shove & N. Spurling (Eds.), *Sustainable practices: Social theory and climate change* (pp. 292–317). London: Routledge.

Sherry, J. F., Jr. (2013). Reflections of a scape artist: Discerning scapus in contemporary worlds. In D. Rinallo, L. M. Scott, & P. Maclaran (Eds.), *Consumption and spirituality* (pp. 211–230). London: Routledge.

Smart, A. (2014). Critical perspectives on multispecies ethnography. *Critique of Anthropology, 34*(1), 3–7.

Smith, B., & Sparkes, A. C. (2005). Men, sport, spinal cord injury, and narratives of hope. *Social Science and Medicine, 61*(5), 1095–1105.

Soron, D. (2010). Sustainability, self-identity and the sociology of consumption. *Sustainable Development, 18*(3), 172–181.

Squire, C. (2012). Narratives and the gift of the future. *Narrative Works: Issues, Investigations and Interventions, 2*, 67–82.

Tsing, A. (2012). Unruly edges: Mushrooms as companion species. *Environmental Humanities, 1*, 141–154.

Uzzell, D., & Räthzel, N. (2009). Transforming environmental psychology. *Journal of Environmental Psychology, 29*(3), 340–350.

Westerhof, G. J., & Bohlmeijer, E. T. (2012). Life stories and mental health: The role of identification processes in theory and interventions. *Narrative Works: Issues, Investigations and Interventions, 2*, 107–128.

Whitehead, H., & Rendell, L. (2014). *The cultural lives of whales and dolphins.* Chicago: University of Chicago Press.

Zittoun, T., & Gillespie, A. (2015). *Imagination in human and cultural development.* London: Routledge.

References

Adams, M. (2014a). Approaching nature, 'sustainability' and ecological crises from a critical social psychological perspective. *Social and Personality Psychology Compass, 8,* 251–262.

Adams, M. (2014b). Inaction and environmental crisis: Narrative, defence mechanisms and the social organisation of denial. *Psychoanalysis, Culture & Society, 19,* 52–71.

Adams, M. (in press). Critical environmental psychology. In B. Gough (Ed.), *Handbook of critical social psychology.* Basingstoke: Palgrave.

Adger, W. N., Barnett, J., Brown, K., Marshall, N., & O'Brien, K. (2013). Cultural dimensions of climate change impacts and adaptation. *Nature Climate Change, 3,* 112–117.

Agar, N. (2001). *Life's intrinsic value: Science, ethics, and nature.* Columbia: Columbia University Press.

Alkon, A. H., & Traugot, M. (2008). Place matters, but how? Rural identity, environmental decision making, and the social construction of place. *City and Community, 7*(2), 97–112.

American Psychologist. (2011, May–June). *Special Issue: Psychology and Global Climate Change, 66*(4), 241–328.

© The Author(s) 2016
M. Adams, *Ecological Crisis, Sustainability and the Psychosocial Subject,*
DOI 10.1057/978-1-137-35160-9

Anderson, K. (2009). Academic perspectives on climate change and social justice. In K. Krish (Ed.), *ESRC seminar series: Mapping the public policy landscape 'How will climate change affect people in the UK and how can we best develop an equitable response?': Joseph Rowntree Foundation* (pp. 5–6). Swindon: Joseph Rowntree Foundation, Local Government Association, ESRC.

Anderson, K. (2013). Why carbon prices can't deliver the 2°C target. Kevinanderson.info.   http://kevinanderson.info/blog/why-carbon-prices-cant-deliver-the-2c-target/. Accessed 18 Dec 2015.

Arnold, L. (1992). *Windscale, 1957: Anatomy of a nuclear accident.* New York: St. Martin's Press.

Barros, V. R., Field, C. B., Dokken, D. J., Mastrandrea, M. D., Mach, K. J., Bilir, T. E., Chatterjee, M., Ebi, K. L., Estrada, Y. O., Genova, R. C., Girma, B., Kissel, E. S., Levy, A. N., MacCracken, S., Mastrandrea, P. R., & White, L. L. (Eds.). (2014). *Climate change 2014: Impacts, adaptation, and vulnerability. Part B: Regional aspects. Contribution of Working Group II to the fifth assessment report of the Intergovernmental Panel on Climate Change []*. Cambridge, UK/New York: Cambridge University Press. 688 pp.

Basel Action Network. (2002). *Exporting harm: The high-tech trashing of Asia.* http://www.ban.org/E-waste/technotrashfinalcomp.pdf. Accessed 18 Dec 2015.

Bastian, B., Loughnan, S., Haslam, N., & Radke, H. R. M. (2012a). Don't mind meat? The denial of mind to animals used for human consumption. *Personality and Social Psychology Bulletin, 38*(2), 247–256.

Bastian, B., Costello, K., Loughnan, S., & Hodson, G. (2012b). When closing the human-animal divide expands moral concern: The importance of framing. *Social Psychological and Personality Science, 3*(4), 421–429.

Beck, U. (2010). Climate for change, or how to create a green modernity? *Theory, Culture and Society, 27*, 254–266.

Behavioural Insights Team. (2014). EAST: Four simple ways to apply behavioural insights. www.behaviouralinsights.co.uk/. Accessed 18 Dec 2015.

Belk, R. W., Wallendorf, M., & Sherry, J. F., Jr. (1989). The sacred and the profane in consumer behavior: Theodicy on the odyssey. *Journal of Consumer Research, 16*(1), 1–38.

Ben-Asher, S., & Goren, N. (2006). Projective identification as a defense mechanism when facing the threat of an ecological hazard. *Psychoanalysis, Culture and Society, 11*, 17–35.

Bilewicz, M., Imhoff, R., & Drogosz, M. (2011). The humanity of what we eat: Conceptions of human uniqueness among vegetarians and omnivores. *European Journal of Social Psychology, 41*, 201–209.

BirdLife International. (2013a). *BirdLife has grown into the largest global conservation partnership.* Presented as part of the BirdLife State of the world's birds website. http://www.birdlife.org/datazone/sowb/casestudy/550. Accessed 18 Dec 2015.

Blake, D. E. (2001). Contextual effects on environmental attitudes and behaviour. *Environment and Behaviour, 33,* 708–725.

Bokova, I. (2013). Preface. In ISSC and UNESCO, *World social science report 2013, changing global environments* (pp. 3–5). Paris: OECD Publishing and UNESCO Publishing.

Borg, K. L. (1999). The "chauffeur problem" in the early auto era: Structuration theory and the users of technology. *Technology and Culture, 40*(4), 797–832.

Boykoff, M. T. (2011). *Who speaks for the climate? Making sense of media reporting on climate change.* Cambridge: Cambridge University Press.

Brewer, P. R. (2012). Polarisation in the USA: Climate change, party politics, and public opinion in the Obama era. *European Political Science, 11*(1), 7–17.

Brierley, C., Haugvaldstad, A., Lotto, B., Michie, S., Shipworth, M., & Tuckett, D. (2014). *Time for change? Climate science reconsidered: Report of the UCL Policy Commission on Communicating Climate Science, 2014.* London: UCL.

Capaldi, C. A., Dopko, R. L., & Zelenski, J. M. (2014). The relationship between nature connectedness and happiness: A meta-analysis. *Frontiers in Psychology, 5,* 976.

Carrington, D. (2015, February 12). Fracking will be allowed under national parks, UK decides. *The Guardian.* http://www.theguardian.com/environment/2015/feb/12/fracking-will-be-allowed-under-national-parks. Accessed 18 Dec 2015.

Castree, N. (2015, March 9). The 'three cultures' problem in global change research. *EnviroSociety.* http://www.envirosociety.org/2015/03/the-three-cultures-problem-in-global-change-research. Accessed 18 Dec 2015.

Cervinka, R., Röderer, K., & Hefler, E. (2011). Are nature lovers happy? On various indicators of well-being and connectedness with nature. *Journal of Health Psychology, 17*(3), 379–388.

Chatzidakis, A. (2015). Guilt and ethical choice in consumption: A psychoanalytic perspective. *Marketing Theory, 15*(1), 79–93.

Choices, D. A. (2015). The robust relationship between conspiracism and denial of (climate) science. *Psychological Science, 26*(5), 667–670.

Christakis, N. A., & Fowler, J. H. (2009). *Connected: The surprising power of our social networks and how they shape our lives.* New York: Little, Brown.

ClimateFocus. (2014, December 18). Clearing tropical rainforests distorts Earth's wind and water systems, packs climate wallop beyond carbon. *ScienceDaily*. www.sciencedaily.com/releases/2014/12/141218080823.htm. Accessed 18 Dec 2015.

Collins, J., Thomas, G., Willis, R., & Wilsdon, J. (2003). Carrots, sticks and sermons: Influencing public behaviour for environmental goals. London: Demos/Green Alliance/Defra. http://www.demos.co.uk/files/CarrotsSticks Sermons.pdf. Accessed 18 Dec 2015.

Cook, B. I., Ault, T. R., & Smerdon, J. E. (2015). Unprecedented 21st century drought risk in the American Southwest and Central Plains. *Science Advances, 1*(1), e1400082. doi:10.1126/sciadv.1400082.

Corner, A. (2015). *Climate silence (and how to break it)*. Climate outreach and information network. http://climateoutreach.org/resources/climate-silence-and-how-to-break-it/. Accessed 18 Dec 2015.

Cortazzi, M. (2001). Narrative analysis in ethnography. In P. Atkinson, A. Coffey, S. Delamont, J. Lofland, & L. Lofland (Eds.), *Handbook of ethnography* (pp. 384–393). London: Sage.

Coughlan, H., Tiedt, L., Clarke, M., Kelleher, I., Tabish, J., Molloy, C., ... & Cannon, M. (2014). Prevalence of DSM-IV mental disorders, deliberate self-harm and suicidal ideation in early adolescence: An Irish population-based study. *Journal of Adolescence, 37*(1), 1–9.

Cowan, E. (1968). *Oil and water – The Torrey Canyon disaster*. Philadelphia: Lippincott.

Cox, R. (2007). Nature's "crisis disciplines": Does environmental communication have an ethical duty? *Environmental Communication, 1*, 5–20.

Craib, I. (2003). The unhealthy underside of narratives. In *Narrative, memory and health* (pp. 1–11). Huddersfield: University of Huddersfield.

Croker, R. (2010, January 28). No garden to get back to: Understanding post-Avatar ecological depressive disorder. *Religion Dispatches*. http://www.religiondispatches.org/archive/culture/2226/nogardentogetbacktounderstandingpostavatarecologicaldepressivedisorder. Accessed 18 Dec 2015.

Davis, M. (2010, January/February). Who will build the Ark? *New Left Review, 61*, 29–46.

Delingpole, J. (2012, March 31). Global Weirding: The new Big Lie. *The Daily Telegraph*. http://blogs.telegraph.co.uk/news/jamesdelingpole/100148381/global-weirding-the-new-big-lie/

Department of Energy and Climate Change. (2014). 2013 UK Greenhouse Gas Emissions, Provisional Figures and 2012 UK Greenhouse Gas Emissions,

Final Figures by Fuel Type and End-User. London: HM Government/ National Statistics. Available https://www.gov.uk/government/publications/ final-uk-emissions-estimates

Dobson, A. (2007). *Green political thought* (4th ed.). London: Routledge.

Dolan, P., Hallsworth, M., Halpern, D., King, D., & Vlaev, I. (2010). *Mindspace*. London: Cabinet Office/Institute for Government. http://www.institutefor-government.org.uk/publications/mindspace. Accessed 18 Dec 2015.

Donald, R. (2012, April 3). Lessons in Newspeak: How to make a sociologist sound Orwellian. *The Carbon Brief*. http://www.carbonbrief.org/ blog/2012/04/are-skeptics-sick/

Dunlap, R. E. (2008). The new environmental paradigm scale: From marginality to worldwide use. *Journal of Environmental Education, 40*, 3–18.

Dunlap, R. E., & McCright, A. M. (2008). A widening gap: Republican and democratic views on climate change. *Environment: Science and Policy for Sustainable Development, 50*(5), 26–35.

Dunn, R. R., Gavin, M. C., Sanchez, M. C., & Solomon, J. N. (2006). The pigeon paradox: Dependence of global conservation on urban nature. *Conservation Biology, 20*, 1814–1816.

Edenhofer, O., Pichs-Madruga, R., Sokona, Y., Farahani, E., Kadner, S., Seyboth, K., Adler, A., Baum, I., Brunner, S., Eickemeier, P., Kriemann, B., Savolainen, J., Schlömer, S., von Stechow, C., Zwickel, T., & Minx, J. C. (Eds.) (2011). *IPCC special report on renewable energy sources and climate change mitigation*. Prepared by Working Group III of the Intergovernmental Panel on Climate Change. Cambridge University Press, Cambridge, UK/ New York, 1075 pp.

ESRC. (2015a). *Public belief in climate change reaches 10 year high*. http://www. esrc.ac.uk/news-and-events/features-casestudies/features/33516/Public_ belief_in_climate_change_reaches_10year_high.aspx. Accessed 22 Dec 2015.

Field, C. B., Barros, V., Stocker, T. F., Qin, D., Dokken, D. J., Ebi, K. L., Mastrandrea, M. D., Mach, K. J., Plattner, G.-K., Allen, S. K., Tignor, M., & Midgley, P. M. (Eds.) (2012). *Managing the risks of extreme events and disasters to advance climate change adaptation*. A special report of Working Groups I and II of the Intergovernmental Panel on Climate Change. Cambridge University Press, Cambridge, UK/New York, 582 pp.

Fivush, R., & Nelson, K. (2004). Culture and language in the emergence of autobiographical memory. *Psychological Science, 15*, 586–590.

Fivush, R., Haden, C. A., & Reese, E. (1996). Remembering, recounting and rem- iniscing: The development of autobiographical memory in social context. In D. Rubin (Ed.), *Reconstructing our past: An overview of autobiographical memory* (pp. 341–359). Cambridge: Cambridge University Press.

Floyd, D. L., Prentice-Dunn, S., & Rogers, R. W. (2000). A meta-analysis of research on protection motivation theory. *Journal of Applied Social Psychology, 30*(2), 407–429.

Foderado, L. W. (2014, September 21). Taking a call for climate change to the streets. *New York Times.* http://www.nytimes.com/2014/09/22/nyregion/new-york-city-climate-change-march.html?_r=1

Fossey, D. (2000). *Gorillas in the mist.* Houghton: Mifflin Harcourt.

Fossil Free. (2015). Global Divestment Day. http://gofossilfree.org/uk/global-divestment-day/. Accessed 18 Dec 2015.

Foster, C., McMeekin, A., & Mylan, J. (2012). The entanglement of consumer expectations and eco-innovation pathways: The case of orange juice. *Technology Analysis and Strategic Management, 24*(4), 391–405.

Friederichs, K. (1958). A definition of ecology and some thoughts about basic concepts. *Ecology, 39*(1), 154–159.

Friedrichs, J. (2013). *The future is not what it used to be: Climate change and energy scarcity.* Cambridge: MIT Press.

Giddens, A. (1984). *The constitution of society. Outline of the theory of structuration.* Cambridge: Polity Press.

Gifford, R., & Nilsson, A. (2014). Personal and social factors that influence pro-environmental concern and behaviour: A review. *International Journal of Psychology, 49*, 141–157.

Globescan. (2014). Increased fears about environment, but little change in consumer behavior, according to new National Geographic/Globescan study. http://www.globescan.com/news-and-analysis/press-releases/press-releases-2014/99-press-releases-2014/328-increased-fears-about-environment-but-little-change-in-consumer-behavior-national-geographic-globescan-study.html. Accessed 18 Dec 2015.

Goffman, E. (1974). *Frame analysis: An essay on the organization of experience.* New York: Harper and Row.

Goldenberg, J. L., & Arndt, J. (2008). The implications of death for health: A terror management health model for behavioral health promotion. *Psychological Review, 115*, 1032–1053.

Goldenberg, J., Mazursky, D., & Solomon, S. (1999). The fundamental templates of quality ads. *Marketing Science, 18*, 333–351.

Goodall, J. (1986). *The chimpanzees of Gombe: Patterns of behavior.* Harvard: Belknap Press of Harvard University Press.

Gorte, R. W., & Shiekh, P. A. (2010). *Deforestation and climate change.* Congressional Research Service. http://forestindustries.eu/sites/default/files/userfiles/1file/R41144.pdf. Accessed 18 Dec 2015.

Green Light. (2015, February 13). Green news roundup: Mega droughts, geo-engineering and straw homes. *The Guardian.* http://www.theguardian.com/environment/2015/feb/13/green-news-roundup-mega-droughts-geoengineering-and-straw-homes?CMP=EMCENVEML1631. Accessed 18 Dec 2015.

Groves, C., Henwood, K., Shirani, F., Butler, C., Parkhill, K., & Pidgeon, N. (2016b). Invested in unsustainability? On the psychosocial patterning of engagement in practices. *Environmental Values, 25*(3), 309–328.

Guber, D. L. (2012). A cooling climate for change? Party polarization and the politics of global warming. *American Behavioral Scientist, 57*(1), 93–115.

Harden, C. P., Chin, A., English, M. R., Fu, R., Galvin, K. A., Gerlak, A. K., & Wohl, E. E. (2014). Understanding human–landscape interactions in the "Anthropocene". *Environmental Management, 53*(1), 4–13.

Hargreaves, T., Nye, M., & Burgess, J. (2008). Social experiments in sustainable consumption: An evidence-based approach with potential for engaging low-income communities. *Local Environment, 13*(8), 743–758.

Hargreaves, D. (2011a). Pro-environmental interaction: Engaging Goff man on pro-environmental behaviour change. *Centre for Social and Economic Research on the Global Environment Working Papers 11-04.* Norwich: University of East Anglia.http://www.cserge.ac.uk/sites/default/files/2011-04.pdf. Accessed 18 Dec 2015.

Hawcroft, L. J., & Milfont, T. L. (2010). The use (and abuse) of the new environmental paradigm scale over the last 20 years: A meta-analysis. *Journal of Environmental Psychology, 30*, 143–158.

Hayes, C. (2014, May 12). The new abolitionism. *The Nation.* http://www.the-nation.com/article/179461/new-abolitionism

Hayes, J., Schimel, J., Arndt, J., & Faucher, E. H. (2010). A theoretical and empirical review of the death-thought accessibility concept in terror management research. *Psychological Bulletin, 136*(5), 699.

Heilbrun, C. (1988). *Writing a woman's life.* New York: Norton.

Hobaiter, C., & Byrne, R. W. (2014). The meanings of chimpanzee gestures. *Current Biology, 24*(14), 1596–1600.

Hobson, K. (2003). Thinking habits into action: The role of knowledge and process in questioning household consumption practices. *Local Environment, 8*(1), 95–112.

Holthaus, J. (2015, July 29). Bug-out scenarios. How do climate scientists cope with existential dread? *Slate.* http://www.slate.com/articles/health_and_science/science/2015/07/climate_scientists_despair_most_devastating_parts_of_esquire_s_jason_box.html. Accessed 18 Dec 2015.

Hornsey, M. J., Fielding, K. S., McStay, R., Reser, J. P., Bradley, G. L., & Greenaway, K. H. (2015). Evidence for motivated control: Understanding the paradoxical link between threat and efficacy beliefs about climate change. *Journal of Environmental Psychology, 42,* 57–65.

Howell, R. A. (2014). It's not (just) the environment, stupid! Values, motivations, and routes to engagement of people adopting lower-carbon lifestyles. *Global Environmental Change, 23*(1), 281–290.

Howell, A. J., Dopko, R. L., Passmore, H. A., & Buro, K. (2011). Nature connectedness: Associations with well-being and mindfulness. *Personality and Individual Differences, 51*(2), 166–171.

Hulme, M. (2008). The conquering of climate: Discourses of fear and their dissolution. *The Geographical Journal, 174*(1), 5–16.

IPCC. (2013). *IPCC Factsheet: What is the IPCC?* http://www.ipcc.ch/news_and_events/docs/factsheets/FS_what_ipcc.pdf. Accessed 18 Dec 2015.

IPPC. (2014a). *Fifth assessment report (AR5).* http://www.ipcc.ch/

IPPC. (2014b). *Fifth assessment report (AR5): Summary for policy makers.* http://www.ipcc.ch/pdf/assessment-report/ar5/syr/SYRAR5SPMcorr2.pdf

IPCC. (2015). *Working groups/task force.* https://www.ipcc.ch/workinggroups/workinggroups.shtml. Accessed 18 Dec 2015.

Jambeck, J. R., Geyer, R. Wilcox, C., Siegler, T. R., Perryman, M., Andrady, A., Narayan, R., & Law, K. L. (2015, February 13). Plastic waste inputs from land into the ocean. *Science, 347*(6223), 768–771.

Josselson, R., & Lieblich, A. (Eds.). (1993). *The narrative study of lives.* Thousand Oaks: Sage.

Kellert, S. R. (1985). Social and perceptual factors in endangered species management. *Journal of Wildlife Management, 49,* 528–536.

Kerry, J. (2014, November 2). *Press statement release of the synthesis report of the fifth assessment of the Intergovernmental Panel on Climate Change.* US Department of State. http://www.state.gov/secretary/remarks/2014/11/233627.htm. Accessed 18 Dec 2015.

Kidner, D. (2007). Depression and the natural world: Towards a critical ecology of psychological distress. *The International Journal of Critical Psychology, 19,* 123–146.

Kingsnorth, P., & Hine, D. (2009). The Uncivilization Manifesto. http://darkmountain.net/about/manifesto/. Accessed 21 Dec 2015.

Kinnvall, C. (2004). Globalization and religious nationalism: Self, identity, and the search for ontological security. *Political Psychology, 25*(5), 741–767.

Kleinman, P. (2012). *Psych 101: Psychology facts, basics, statistics, tests, and more!* Adams Media.

Kobayashi, A., & Mackenzie, S. (2014). *Remaking human geography.* London: Routledge.

Kurukulasuriya, P., & Rosenthal, S. (2013) *Climate change and agriculture : A review of impacts and adaptations.* Washington, DC: World Bank. https://openknowledge.worldbank.com/handle/10986/16616. Accessed 18 Dec 2015.

Laing, R. D. (1967). *The politics of experience and the bird of paradise.* Harmondsworth: Penguin.

Laing, R. D., & Esterson, A. (1964). *Sanity, madness and the family.* London: Penguin.

Langenheim, J. (2015, February 12). Indonesia winning battle to save world's richest reef system. *The Guardian.* http://www.theguardian.com/environment/the-coral-triangle/2015/feb/12/indonesia-winning-battle-to-save-worlds-richest-reef-system. Accessed 18 Dec 2015.

Lawler, S. (2005). Disgusted subjects: The making of middle-class identities. *The Sociological Review, 53*(3), 429–446.

Lawrence, D., & Vandecar, K. (2014). Effects of tropical deforestation on climate and agriculture. *Nature Climate Change, 5*(1), 27.

Lele, S. M. (1991). Sustainable development: A critical review. *World Development, 19*(6), 607–621.

Lenton, T. M. (1998). Gaia and natural selection. *Nature, 394*(6692), 439–447.

Lertzman, R. (2004). Ecopsychological theory and critical intervention. *Organization and Environment, 17*(3), 396.

Limbaugh, R. (2012, April 2). Environmentalist wacko: Climate change skeptics are sick. http://www.rushlimbaugh.com/daily/2012/04/02/environmentalist_wacko_climate_change_skeptics_are_sick. Accessed 18 Dec 2015.

Lovelock, J. E. (1965). A physical basis for life detection experiments. *Nature, 207*(4997), 568–570.

Lowe, T., Brown, K., Dessai, S., de França Doria, M., Haynes, K., & Vincent, K. (2006). Does tomorrow ever come? Disaster narrative and public perceptions of climate change. *Public Understanding of Science, 15*(4), 435–457.

Luokkanen, M., Huttunen, S., & Hildén, M. (2014). Geoengineering, news media and metaphors: Framing the controversial. *Public Understanding of Science, 23*(8), 966–981.

Maheswaran, D., & Agrawal, N. (2004). Motivational and cultural variations in mortality salience effects: Contemplations on terror management theory and consumer behavior. *Journal of Consumer Psychology, 14*(3), 213–218.

Mail Online. (2012, March 31). If you don't believe in climate change you must be sick': Oregon professor likens skepticism to racism. http://www.dailymail.co.uk/news/article-2123260/If-dont-believe-climate-change-sick-Oregon-professor-likens-skepticism-racism.html. Accessed 18 Dec 2015.

Maina, J., de Moel, H., Zinke, J., Madin, J., McClanahan, T., & Vermaat, J. E. (2013). Human deforestation outweighs future climate change impacts of sedimentation on coral reefs. *Nature Communications, 4*(1986): 1–7.

Mair, P., & Van Biezen, I. (2001). Party membership in twenty European democracies, 1980–2000. *Party Politics, 7*(1), 5–21.

Margulis, L. (2004). Gaia by any other name. In S. Schneider, J. Miller, E. Crist, & P. Boston (Eds.), *Scientists debate Gaia: The next century* (pp. 7–12). Cambridge, MA: MIT Press.

Matthews, F. (2003). Living with animals. In S. J. Armstrong (Ed.), *The animal ethics reader*. Hove: Psychology Press.

Matthews, L., & Matthews, A. (2015, February 23). Five ways that people frame climate change debates. *The Guardian.* http://www.theguardian.com/sustainable-business/2015/feb/23/five-ways-that-people-frame-climate-change-debates. Accessed 18 Dec 2015.

McAdams, D. P. (2001). Generativity in midlife. In M. Lachman (Ed.), *Handbook of midlife development* (pp. 395–443). New York: Wiley.

McCright, A. M., & Dunlap, R. E. (2011). The politicization of climate change and polarization in the American public's views of global warming, 2001–2010. *The Sociological Quarterly, 52*(2), 155–194.

McGrath, M. (2013, October 23). UN climate chief's tears over future generations. *BBC News.* http://www.bbc.co.uk/news/science-environment-24615946. Accessed 18 Dec 2015.

McKibben, B. (2015, March 9). Pressure is growing. *The Guardian.* http://www.theguardian.com/environment/2015/mar/09/climate-fight-wont-wait-for-paris-vive-la-resistance. Accessed 18 Dec 2015.

Merskin, D. (2010). Hearing voices: The promise of participatory action research for animals. *Action Research, 9*(2), 144–161.

Millennium Ecosystem Assessment. (2005). *Ecosystems and human well-being: Synthesis*. Washington, DC: Island Press.

Miyashita, A. (2007). Where do norms come from? Foundations of Japan's post-war pacifism. *International Relations of the Asia–Pacific, 7*, 99–120.

Moss, R. H., Edmonds, J. A., Hibbard, K. A., Manning, M. R., Rose, S. K., Van Vuuren, D. P., & Wilbanks, T. J. (2010). The next generation of scenarios for climate change research and assessment. *Nature, 463*(7282), 747–756.

Mouhot, J. F. (2011). Past connections and present similarities in slave owner-ship and fossil fuel usage. *Climatic Change, 105*(1–2), 329–355.

Nash, J. E., & Sutherland, A. (1991). The moral elevation of animals: The case of "gorillas in the mist". *International Journal of Politics, Culture, and Society, 5*(1), 111–126.

National Research Council. (2015). *Climate intervention: Carbon dioxide removal and reliable sequestration*. Washington, DC: National Academies Press.

Newhouse, N. (1990). Implications of attitude and behavior research for envi-ronmental conservation. *The Journal of Environmental Education, 22*(1), 26–32.

Nisbet, M. C. (2009). Communicating climate change: Why frames matter for public engagement. *Environment: Science and Policy for Sustainable Development, 51*(2), 12–23.

Nisbet, E. K., Zelenski, J. M., & Murphy, S. A. (2009). The nature relatedness scale: Linking individuals' connection with nature to environmental concern and behavior. *Environment and Behavior, 41*(5), 715–740.

Nobre, C. A., Sellers, P. J., & Shukla, J. (1991). Amazonian deforestation and regional climate change. *Climate, 4*, 957–988.

Nolt, J. (2011). Nonanthropocentric climate ethics. *Wiley Interdisciplinary Reviews: Climate Change, 2*, 701–711.

O'Neill, S., & Nicholson-Cole, S. (2009). 'Fear won't do it': Promoting positive engagement with climate change through visual and iconic representations. *Science Communication, 30*(3), 355–379.

Page, L. (2012, March 30). Climate-change scepticism must be 'treated', says enviro-sociologist. *The Register*. http://www.theregister.co.uk/2012/03/30/climate_scepticism_racism_slavery_treatment/. Accessed 18 Dec 2015.

Palmer, C. (2011). Does nature matter? The place of the nonhuman in the ethics of climate change. In D. G. Arnold (Ed.), *The ethics of global climate change* (pp. 272–279). Cambridge: Cambridge University Press.

Palomo, I., Montes, C., Martín-López, B., González, J. A., García-Llorente, M., Alcorlo, P., & Mora, M. R. G. (2014). Incorporating the social–ecological approach in protected areas in the Anthropocene. *BioScience*, bit033.

Pascale, C. M. (2010). *Cartographies of knowledge: Exploring qualitative epistemologies*. London: Sage.

Peterson Del Mar, D. (2014). *Environmentalism*. London: Routledge.

Princeton University. (2015). *Man's role in changing the face of the earth* (1955). http://www.princeton.edu/forbescollege/about/history/(mans-role-in-changing-th/. Accessed 21 Dec 2015.

Pyszczynski, T., Solomon, S., & Greenberg, J. (2015). Thirty years of terror management theory: From genesis to revelation. *Advances in Experimental Social Psychology, 52*, 1–70.

Pyszczynslu, T., Solomon, S., & Greenberg, J. (2003). *In the wake of 9/11: The psychology of terror*. Washington, DC: American Psychological Association.

Quinn, J., Pascoe, A., Wood, W., & Neal, D. (2010). Can't control yourself? Monitor those bad habits. *Personality and Social Psychology Bulletin, 36*(4), 499–511.

Randles, S., & Mander, S., (2009). Practice(s) and Ratchet(s): A sociological examination of frequent flying. In S. Gossling & P. Upham (Eds.), *Climate Change and aviation: Issues, challenges and solutions*. London: Earthscan.

Rapley, C. G., de Meyer, K., Carney, J., Clarke, R., Howarth, C., Smith, N., Stilgoe, J., Youngs, S., Reid, W. V., Chen, D., Goldfarb, L., Hackmann, H., Lee, Y. T., Mokhele, K., & Whyte, A. (2010). Earth system science for global sustainability: Grand challenges. *Science, 330*(6006), 916–917.

Readfearn, G. (2015, February 12). Climate science denialists in tailspin over hottest years. *The Guardian*. http://www.theguardian.com/environment/planet-oz/2015/feb/12/climate-science-denialists-in-tailspin-over-hottest-years. Accessed 18 Dec 2015.

Respaut, R. (2014, December 16). California pensions should divest coal assets – State senate leader. *Reuters*. http://uk.reuters.com/article/2014/12/16/california-coal-pensions-idUKL1N0U01Y620141216. Accessed 21 Dec 2015.

Roepstorff, A. (2001). Thinking with animals. *Sign Systems Studies, 1*, 203–218.

Roszak, T. (2009). A psyche as big as the earth. In L. Buzzell & C. Chalquist (Eds.) (2010), *Ecotherapy: Healing with nature in mind*. San Francisco, CA: Sierra Club Books.

Rowson, J., & Corner, A. (2015). *The seven dimensions of climate change: Introducing a new way to think and talk and act*. London: RSA/COIN. https://www.thersa.org/discover/publications-and-articles/reports/the-seven-dimensions-of-climate-change-introducing-a-new-way-to-think-talk-and-act/. Accessed 18 Dec 2015.

Sayer, A. (2005). Class, moral worth and recognition. *Sociology, 39*(5), 947–963.

Sayers, S. (1998). *Marxism and human nature.* Hove: Psychology Press.

Schatzki, T. R. (1996). *Social Practices: A Wittgensteinian approach to human activity and the social.* Cambridge: Cambridge University Press.

Schmidt, A., Ivanova, A., & Schäfer, M. S. (2013). Media attention for climate change around the world: A comparative analysis of newspaper coverage in 27 countries. *Global Environmental Change, 23*(5), 1233–1248.

Schumacher, I. (2014). An empirical study of the determinants of green party voting. *Ecological Economics, 105,* 306–318.

Scruggs, L., & Benegal, S. (2012). Declining public concern about climate change: Can we blame the great recession? *Global Environmental Change, 22*(2), 505–515.

Selinger, E., & Whyte, K. (2011). Is there a right way to nudge? The practice and ethics of choice architecture. *Sociology Compass, 5*(10), 923–935.

Shanahan, J., Pelstring, L., & McComas, K. (1999). Using narratives to think about environmental attitude and behavior: An exploratory study. *Society and Natural Resources, 12*(5), 405–419.

Sherif, M., Harvey, O. J., White, B. J., Hood, W. R., & Sherif, C. W. (1961). *Intergroup conflict and co-operation: The Robbers Cave experiment.* Norman: University of Oklahoma.

Shove, E. (2003). *Comfort, cleanliness and convenience: The social organization of normality.* Oxford: Berg.

Shove, E. A., & Pantzar, M. (2007). Recruitment and reproduction: The careers and carriers of digital photography and floorball. *Journal of Human Affairs, 17,* 154–167.

Shove, E. A., & Walker, G. (2010). Governing transitions in the sustainability of everyday life. *Research Policy, 39,* 471–476.

Shove, E. A., Watson, M., Hand, M., & Ingram, J. (2007). *The design of everyday life.* Oxford: Berg.

Shukla, J., Nobre, C., & Sellers, P. (1990). Amazon deforestation and climate change. *Science, 247,* 1322–1325.

Skeggs, B. (2005). The making of class and gender through visualizing moral subject formation. *Sociology, 39*(5), 965–982.

Smelser, N. (2004). Psychological trauma and cultural trauma. In J. Alexander, R. Eyerman, B. Giesen, N. Smelser, & P. Sztompka (Eds.), *Cultural trauma and collective identity* (pp. 31–59). Oakland: University of California Press.

Smith, N., & Leiserowitz, A. (2012). The rise of global warming skepticism: Exploring affective image associations in the United States over time. *Risk Analysis, 32*, 1021–1032.

Solomon, S., Greenberg, J., Pyszczynski, T., Cohen, F., & Ogilvie, D. (2009). Teach these souls to fly: Supernatural as human adaptation. In *Evolution, culture, and the human mind* (pp. 99–118). New York: Psychology Press.

Sools, A. (2012). "To see a world in a grain of sand": Towards future-oriented what-if analysis in narrative research. *Narrative Works, 2*(1), 83.

Southerton, D., Mcmeekin, A., & Evans, D. (2011). *International review of behaviour change initiatives: Climate Change behaviours research programme.* Edinburgh: Scottish Government Social Research.

Stocker, T. F., Qin, D., Plattner, G.-K., Tignor, M., Allen, S. K., Boschung, J., Nauels, A., Xia, Y., Bex, V., & Midgley, P. M. (Eds.) (2013b). *Climate change 2013: The physical science basis. Contribution of Working Group I to the Fifth assessment report of the Intergovernmental Panel on Climate Change* [Stocker, T. F., Qin, D., Plattner, G.-K., Tignor, M., Allen, S. K., Boschung, J., Nauels, A., Xia, Y., Bex, V., & Midgley, P. M. (Eds.)]. Cambridge, UK/New York: Cambridge University Press, 1535 pp.

Thaler, R., & Sunstein, C. (2008). *Nudge: Improving decisions about health, wealth, and happiness.* Harmondsworth: Penguin.

The British Psychological Society's Behaviour Change Advisory Group (BCAG). (2014). *Behaviour change: Energy conservation.* Leicester: British Psychological Society. http://www.bps.org.uk/system/files/Public%20files/energy.pdf. Accessed 18 Dec 2015.

Thomas, W. L., Jr. (Ed.). (1956). *Man's role in changing the face of the earth.* Chicago: The University of Chicago Press.

Trigwell, J. L., Francis, A. J., & Bagot, K. L. (2014). Nature connectedness and eudaimonic well-being: Spirituality as a potential mediator. *Ecopsychology, 6*(4), 241–251.

Turner, B. (1984). *Body and society: Explorations in social theory.* London: Sage.

U.S. Fish and Wildlife Service. (2015). *Save the monarch butterfly.* U.S. Government. http://www.fws.gov/savethemonarch/

UK Climate Change Committee. (2008). *Building a low-carbon economy – The UK's contribution to tackling climate change.* http://www.theccc.org.uk/publication/building-a-low-carbon-economy-the-uks-contribution-to-tackling-climate-change-2/. Accessed 18 Dec 2015.

UNFCCC. (2011). *Conference of the Parties – Sixteenth Session: Decision 1/CP.16: The Cancun Agreements: Outcome of the work of the Ad Hoc Working*

Group on Long-term Cooperative Action under the Convention (English): Paragraph 4 (PDF) (p. 3). UNFCCC Secretariat: Bonn: UNFCCC.

United Nations Environment Programme. (1992). *Agenda 21.* http://www. unep.org/documents/default.asp?documentid=52. Accessed 18 Dec 2015.

University of Oregon. (2012, March 26). *Simultaneous action needed to break cultural inertia in climate-change response.* http://uonews.uoregon.edu/ archive/news-release/2012/3/simultaneous-action-needed-break-cultural-inertia-climate-change-respons. Accessed 18 Dec 2015.

Urry, J. (2013). *Societies beyond oil: Oil dregs and social futures.* London: Zed Books.

Vail, K. E., Rothschild, Z. K., Weise, D. R., Solomon, S., Pyszczynski, T., & Greenberg, J. (2010). A terror management analysis of the psychological functions of religion. *Personality and Social Psychology Review, 14*(1), 84–94.

Vaughn, A. (2015, February 10). UK spent 300 times more on fossil fuels than clean energy despite green pledge. *The Guardian.* http://www.theguardian. com/environment/2015/feb/10/uk-spent-300-times-more-fossil-fuel-clean-energy-despite-green-pledge

Verbong, G. P. J., & Geels, F. W. (2010). Exploring sustainability transitions in the electricity sector with socio-technical pathways. *Technological Forecasting and Social Change, 77*(8), 1214–1221.

Walker, T. (2015, April 22). Chimpanzees held in lab are given legal rights. *The Independent,* p. 18.

Walker, G., & Shove, E. A. (2007). Ambivalence, sustainability and the governance of socio-technical transitions. *Journal of Environment Policy and Planning, 9*(3-4), 213–225.

WCED. (1987). *Our common future.* Oxford: Oxford University Press.

Weber, E. U. (2006). Experience-based and description-based perceptions of long-term risk: Why global warming does not scare us (yet). *Climatic Change, 77*(1–2), 103–120.

Weber, E. U., & Stern, P. C. (2011). Public understanding of climate change in the United States. *American Psychologist, 66*(4), 315–328.

Weingart, P., Engels, A., & Pansegrau, P. (2000). Risks of communication: Discourses on climate change in science, politics, and the mass media. *Public Understanding of Science, 9*(3), 261–283.

Welch, D., & Warde, A. (2015). Theories of practice and sustainable consumption. In L. Reisch & J. Thøgersen (Eds.), *Handbook of research on sustainable consumption* (pp. 84–100). Cheltenham: Edward Elgar Publishing.

Wetherell, M. (2015). Trends in the turn to affect: A social psychological critique. *Body and Society, 21*(2), 139–166.

Whiten, A., Goodall, J., McGrew, W. C., Nishida, T., Reynolds, V., Sugiyama, Y., Tutin, C. E. G., Wrangham, R. W., & Boesch, C. (1999). Cultures in chimpanzees. *Nature, 399*(6737), 682–685.

Whitmarsh, L. (2011). Scepticism and uncertainty about climate change: Dimensions, determinants and change over time. *Global Environmental Change, 21*, 690–700.

Whitmarsh, L. E., O'Neill, S., & Lorenzoni, I. (2013). Public engagement with climate change: What do we know and where do we go from here? *International Journal of Media and Cultural Politics, 9*(1), 7–25.

Winthrop, S. (2014). From global change science to action with social sciences. *Nature Climate Change, 4*(8), 656–659.

Wolf, E. R. (1957). General and theoretical: Man's role in changing the face of the earth: William L. Thomas, Jr. *American Anthropologist, 59*(6), 1089–1091.

World Wildlife Foundation. (2015). *Deforestation and climate change*. http://www.wwf.org.uk/whatwedo/forests/deforestationandclimatechange/. Accessed 18 Dec 2015.

Worldwatch Institute. (2013). *Despite disappointment, climate summit marks high point for activist movement*. http://www.worldwatch.org/node/6355. Accessed 18 Dec 2015.

Yip, K. S. (2006). *Clinical practice for people with schizophrenia: A humanistic and empathetic encounter*. New York: Nova Publishers.

Yusoff, K., & Gabrys, J. (2011). Climate change and the imagination. *WIRES Climate Change, 2*, 516–534.

Index

Note: Page numbers with "n" denote endnotes.

© The Author(s) 2016

M. Adams, *Ecological Crisis, Sustainability and the Psychosocial Subject*,
DOI 10.1057/978-1-137-35160-9

CPI Antony Rowe

Chippenham, UK

2017-02-03 12:07